FISHING FLORIDA BY PADDLE

AN ANGLER'S GUIDE

JOHN KUMISKI

Foreword by Doug Olander,
editor & content director, *Sport Fishing Magazine*

Published by The History Press
Charleston, SC
www.historypress.net

Copyright © 2019 by John Kumiski
All rights reserved

All images appear courtesy of the author unless otherwise noted.

First published 2019

Manufactured in the United States

ISBN 9781467140638

Library of Congress Control Number: 2019945089

Notice: The information in this book is true and complete to the best of our knowledge. It is offered without guarantee on the part of the author or The History Press. The author and The History Press disclaim all liability in connection with the use of this book.

All rights reserved. No part of this book may be reproduced or transmitted in any form whatsoever without prior written permission from the publisher except in the case of brief quotations embodied in critical articles and reviews.

*This book is dedicated to my father,
who infected me with the fishing pox at a very young age.*

The author and his father, John Kumiski. *Photo by Pauline Kumiski.*

CONTENTS

Foreword by Doug Olander	9
Acknowledgements	11
Introduction	13
How to Use This Book	13
Disclaimer	14
On the Author's Biases	14
On the Author's Assumptions	15
On Finding What Works for You	16
Guided Trips vs. Doing It Yourself	16
A Brief History of Paddle Fishing in Florida	17
PART 1: ADVICE	
1. Boats	23
Kayaks, Canoes and Paddleboards	23
2. Accessories	26
Flotation Devices, Required Safety Gear and	
Optional Safety Gear	26
Paddles, Anchors, Stake-Out Poles and Push Poles	28
Aids to Navigation	31
Sun and Weather Protection	33
Wading Accessories	35
On Keeping Your Gear Dry	36
3. Planning	37
Licenses	37
Timing Your Trip	37
Trip Types and Vehicle Security	38

Contents

Boating on Saltwater	39
4. Tackle	40
Spin Reels, Rods, Lines and Leaders: On Lure Choice and Covering the Water Column	40
Fly Reels, Rods, Lines and Leaders: On Fly Choice and Covering the Water Column	43
5. Fishing Techniques	45
Searching for and Finding Fish Inshore and Stalking Fish	45
Presenting the Bait	49
On Sweat Equity	53
Searching for and Finding Fish in Fresh Water	53
Fighting Fish from a Paddle Vessel	53
Handling and Successfully Releasing Fish	55
Seasonal Considerations: How the Seasons Affect Fish and Fishermen	56
Paddle Fishing Safety: Using Your Head to Stay out of Trouble	57
PART 2: THE FISH	
1. Freshwater Fish	63
2. Saltwater Fish	66
PART 3: DESTINATIONS	
1. Panhandle West	73
Destin	73
Panama City	77
Pensacola Area	77
Fort Pickens	78
Pensacola Proper	79
Surrounding Areas	79
Off the Beaches	81
Freshwater Options	82
Blackwater River	82
Coldwater Creek	82
Escambia River	83
Pensacola History	83
2. Panhandle East	85
Alligator Point to St. Marks	85
Apalachicola Bay and St. George Sound	87
Freshwater Fishing	89
St. Joe Bay	90
Indian Pass	93

Contents

3. North Florida	94
Big Bend Paddling Trail	94
A Big Bend Paddle Adventure	94
Bulow Creek and Tomoka River	101
4. Jacksonville Area	103
Clapboard Creek	104
Fort George Inlet	105
Guana River	105
Ocala National Forest	106
Ocklawaha River	106
Pellicer Creek	108
St. Marys River	111
5. Suwannee and Santa Fe Rivers	112
Suwannee River	112
Sante Fe River	114
6. East Central Florida	115
Fellsmere Grade Recreation Area	115
Indian River Lagoon System	118
Upper Section	119
Canaveral National Seashore	119
Turnbull Creek	122
Banana River Lagoon No-Motor Zone	123
Sebastian River	124
Turkey Creek, Goat Creek, Crane's Creek	126
Southern Section	127
Spruce Creek	127
7. The St. Johns River System	132
The Shad Run	132
Econlockhatchee River	134
St. Johns River Spring Creek Tributaries	140
Rock Springs Run	145
8. West Central Florida	146
Anclote Key	146
Chassahowitzka-Homosassa-Crystal River	147
Hillsborough River	150
Little Manatee River	151
Loxahatchee River	152
Manatee River	153
Sarasota Bay	155
Tampa Bay	156
Tenoroc Fish Management Area and Other Phosphate Pit Lakes	159

Contents

Pits in Polk County	160
Pits in Hamilton County	161
Withlacoochee River	162
Rainbow River	163
9. South Florida	165
Charlotte Harbor and Pine Island Sound	165
Ding Darling National Wildlife Refuge	168
Fort Myers Historical Note	168
Everglades National Park	170
Everglades City/Chokoloskee	171
Flamingo	175
Myakka River	178
Deer Prairie Creek	180
Other Charlotte County Paddle Fishing	181
The Myakka River and the Fountain of Youth	182
Peace River	184
Peacock Bass in South Florida	188
PART 4: PADDLE FISHING THE SEA	
1. Paddle Fishing from the Beach	193
2. Kayak Fishing Offshore by Mike Conneen	196
Appendix (Florida Fishing Licenses, For More Information, Gear Lists)	201
Bibliography	205
About the Author	207

FOREWORD

With about 1,200 miles of shoreline and countless lakes and waterways, Florida is truly a paradise for anglers who paddle canoes, kayaks and stand-up paddleboards. Many of those anglers will find John Kumiski's *Fishing Florida by Paddle* indispensable.

John is happiest with a fishing rod in one hand, a paddle in the other and his butt on the seat of his canoe. I know this because I've fished with him. His decades of experience pursuing this activity are revealed in this book.

I read it not so much as the complete guide for all aspects of paddle fishing all of Florida but rather as an interesting and lively compendium full of classic Kumiski anecdotes and experiences and loads of practical tips. And then, there are the places—although I've kayak fished the state for many years, *Fishing Florida by Paddle* reminds me of all the cool places I haven't fished (but need to!).

No matter what one's level of expertise or experience, he or she will find plenty of information and inspiration in *Fishing Florida by Paddle*.

Doug Olander
Editor in chief and content director
Sport Fishing Magazine

ACKNOWLEDGEMENTS

Any author producing a book requires so much help from so many people! This one is no exception.

Susan, my bride, encouraged me to take on this project and periodically "kicked me in the butt" to get it done. My sons Maxx and Alex were frequent fishing partners and photographic models.

Enough good things cannot be said about Mike Conneen. Without him, this book simply would not exist. Paddle fishing partner, adventurer, lover of living, friend—he's accompanied me on trips as short as a few hours and as long as nineteen days. He has maintained grace and good humor through some trying times!

Tammy Wilson deserves some love, too. She taught me to look out of the box for paddle fishing spots and has been a friend and paddling partner for a long time.

Jim Tedesco started paddling with me back in the 1970s up in New England and accompanied me on some of the paddle fishing safaris this book required.

Other paddle fishing partners include Tom Mitzlaff, Scott and Jack Radloff, Ken Shannon, Rodney Smith and Tom Van Horn. I've had great days with and appreciate the friendship of all of them.

Special mention needs to be made of the paddle fishing guides who helped. Chris Gatz in Fort Walton Beach, Dee Kaminski in Melbourne, Nick Lytle in Navarre, Brian Stauffer in Crystal River and Logan Totten in Port Charlotte were all gracious and giving. I recommend all of them without reservation.

Acknowledgements

Two guys who took me by the hand when I first got to Florida so many years ago and shared everything they knew about fishing here were Steve Baker and Ron Rebeck. There's a never-ending debt of gratitude there, and I can only hope I have paid it forward somewhat.

Eric Johnson at the Florida Fish and Wildlife Conservation Commission gave me plenty of ideas about where to go fishing and what to include in this book.

I appreciate Doug Olander's foreword. Thanks, Doug!

My editors at The History Press are professional wordsmiths and teachers without whom, again, this book would not exist. Amanda Irle started the process, then turned me over to Artie Crisp.

The help of so many wonderful people makes me realize how blessed I am. Profound thanks to all of you.

INTRODUCTION

HOW TO USE THIS BOOK

Thank you for purchasing *Fishing Florida by Paddle: An Angler's Guide*! You'll find it to be a fun and informative tool for planning paddle fishing trips throughout the Sunshine State. The book is divided into three sections:

1. Planning your trip: paddle fishing gear, accessories, when to fish Florida, tackle, techniques, safety and seasonal considerations.
2. The fish you find here, including everything from stumpknockers to tarpon.
3. The areas in the state where the hull meets the water and the actual paddling occurs. We haven't tried to cover everywhere you can paddle fish (this couldn't be done!) but rather have handpicked what we believe are the best places in each region. We have also added some of the history of each region that helps give that place its unique "feel."

So, please come along and let me help you find the best paddle fishing Florida has to offer! Tight Lines!

Introduction

DISCLAIMER

This book provides information in regard to the subject matter covered. It cannot substitute for good judgment or common sense. The reader ventures onto or into the water at his or her own risk.

We have made every effort to make this book as complete and as accurate as possible. However, there may be errors both typographical and in content. Therefore, use this book only as a general guide and not as the ultimate source of boating or fishing information.

The purpose of this book is to educate and entertain. The author and The History Press shall have neither liability not responsibility to any person with regard to any loss or damage caused, or alleged to be caused, directly or indirectly, by the information contained in this book. If the reader does not wish to be bound by the above, he or she may return this book to The History Press for a complete refund.

ON THE AUTHOR'S BIASES

The author has biases, oh, yes. He likes beautiful, wild places. He does not like combat fishing. He loves clear, clean water. He does not like fishing in urban areas or in people's yards.

In Florida, you can catch fish at spillways, power plants, overpasses, discharge tubes, seawalls and so forth. I would never tell anyone not to fish at places like this. You'll find fish anywhere the water is clean enough. However, aesthetics matter. I don't like to fish in "improved" areas, and those types of places are not covered in this book. I have avoided discussing big waters simply because small waters play to paddle vessels' strengths in a way big waters cannot.

My preference is for sit-on-top kayaks and canoes over sit-inside kayaks and paddleboards. I've used all types of boats. It's natural to have favorites. Sit-insides and SUPs (stand-up paddleboards) are fine. It's strictly personal preference—you should use what you like best.

In spite of my being a fly-fisher, I am also a minimalist. Simple is good! My paddleboats don't have all the gadgetry often associated with paddle vessels. There's nothing wrong with gadgetry—I just don't use it.

I prefer lures with single hooks. The practical reason for this is that gang hooks are seldom weedless enough to use in many of the places I fish. They

Introduction

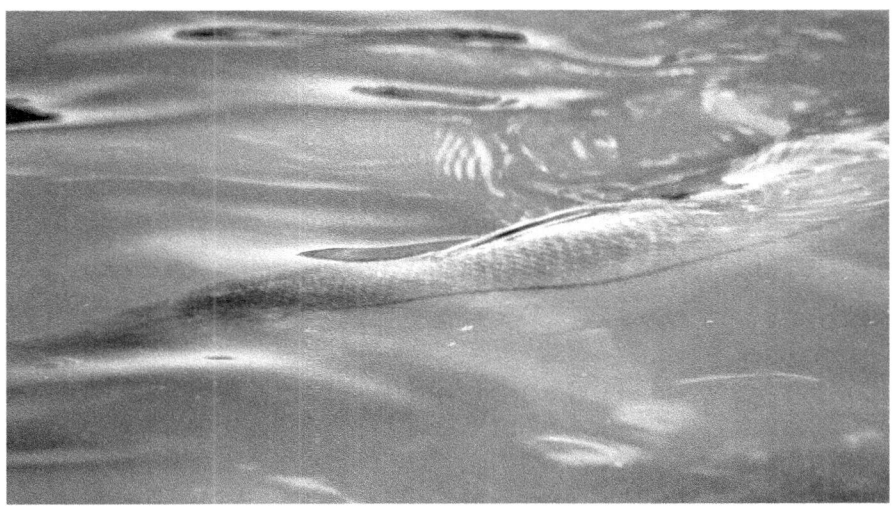

A snook swims through an Everglades pond.

are also safer for both fish and fisherman. But, please, use whatever gives you the most confidence.

I prefer, to a large degree, using artificial lures, especially flies. I dislike using natural bait and never use synthetic bait.

While fishing, I am exceptionally attuned to visual stimuli. I want to see fish, fish activity, signs of fish, signs of life. When these things are not present, I tend not to fish hard or do well in terms of catching. There is much more to paddle fishing than catching fish!

ON THE AUTHOR'S ASSUMPTIONS

This book assumes the reader has achieved a moderate level of experience in fishing. We assume you can tie on a hook, attach a leader, rig a soft plastic bait, set the drag on the reel—that sort of thing. If you don't know how to do those things, you should still read this book. But you should also get a book on fishing for beginners.

We also assume you have a smartphone, tablet or computer that you can use to navigate to the access points to the water as discussed in this book. These devices will get you there much better than a description in a book could.

Introduction

ON FINDING WHAT WORKS FOR YOU

We've discussed my biases. If you ask ten fishing "experts" about the best way to do any particular task (which knot to use to connect line to leader, for example), you're likely to get ten different answers. There's no "best" way to do anything in fishing except as it relates to you. If you're like me, what you consider "best" changes through time, too.

This book explains my current approach to paddle fishing. There are reasons for everything I do, and all of it works perfectly—for me. My methods may or may not work so well for you.

Use this book as a starting point to find your own way. I've been paddle fishing for a long time and can provide helpful insights, but there's no substitute for experience. Don't be afraid to experiment. Find out what works best for you, and do things that way!

GUIDED TRIPS VS. DOING IT YOURSELF

Should you hire a guide or make a go of it on your own? An argument can be made for both approaches.

Hiring a guide significantly increases your costs. However, if you're new to Florida fishing or to the area where you want to fish, a good guide can teach you more in a day than you would learn in months of trying to figure it out on your own. Additionally, a guide will keep you from getting lost. That's good justification for spending the money.

Keep in mind that even the best guides have off days while fishing now and then; you just might hit one.

On the other hand, when you finally do figure it out for yourself, you get a feeling of satisfaction you cannot attain when being guided. Do you have several days to fish? Will you accept a few days of little or no catching in hopes of a payoff at the end?

If you're going to fish the same area for several days, you may want to have someone knowledgeable show you around for a day to kick-start your adventure. Ultimately, the choice of whether to hire a guide or not is yours.

Introduction

A BRIEF HISTORY OF PADDLE FISHING IN FLORIDA

Prior to the arrival of Christopher Columbus, Florida's aboriginal tribes—the Calusa, Tequesta, Ais, Timucua and several others—made a large part of their living by harvesting the rich fish and shellfish resources available here. An analysis of faunal remains at an archaeological site on Sanibel Island showed that more than 93 percent of the energy from animals in the natives' diet came from fish and shellfish, less than 6 percent of the energy came from mammals and less than 1 percent came from birds and reptiles. They may have used traps, spears and nets, but these people were fishermen!

The oldest canoe found in the Western Hemisphere was uncovered on the northern shore of Lake Parker northeast of Lakeland. A radiocarbon analysis indicates this canoe was about three thousand years old. It was in use one thousand years before the birth of Christ! These boats were hand-carved dugout canoes, made mostly from cypress logs, powered by human muscle and propelled with both hand-carved paddles and push poles. Paddle fishing—although not as we know it today—predates the Europeans' arrival in Florida by a couple of millennia.

Something to keep in mind while paddling on Florida's waters is that long before the interstate highway system, automobiles or even bicycles, these waters were the highways of the aboriginal peoples of Florida. The rivers were less altered then and surely had many more fish. The natives moved on them using a method similar to that used by modern paddlers. When you paddle, you're carrying on a tradition that is thousands of years old!

More recently, in the early 1900s, Zane Grey and Ernest Hemingway did a lot to popularize Florida fishing. These men did not paddle, however. The first big game paddle fisherman was most likely Anthony Weston Dimock. His *Book of the Tarpon* was first published in 1911.

A February morning in 1882 found Dimock fishing from a fragile canoe with Tat, his guide and stern man. At the mouth of Florida's Homosassa River, where the current sweeps past Shell Island and out into the Gulf of Mexico, Dimock hooked a tarpon while drifting shrimp for sea trout.

When the fish leaped its first awesome leap, Dimock writes, "The brilliant rays of the semi-tropical sun made a prism of every drop in the shower that surrounded the creature….At first I thought the wonderful being was a mermaid, as I noted her fierce display of activity and strength. I pitied the merman who came home late….Then I suspected it was a wicked genie freed from the Seal of Solomon, which had imprisoned it for thousands of years.

Introduction

"I was brought back to earth by Tat: 'Mus' be a tarpum!'

"'What's that!' I asked.

"'That's what got your hook.'

"Talking in circles is profitless and I turned to my buzzing reel, shouting as I saw the diminishing line: 'Pull like smoke, Tat! Line's 'most gone.'

"Then I put on the drag, but it had no effect. I held my rod vertically and pressed my thumb hard on the reel. Once more the creature shot high in the air while my thumb got red hot. This was in February, 1882, three years before the recognition of the tarpon as a game fish. I believe the tarpon then on my line is entitled to the credit of being the first of its species captured with rod and reel."

Dimock spent thousands of hours exploring Florida's Gulf Coast by motor launch and canoe, searching and fishing for tarpon from Cedar Key down into the Everglades and Keys and even up the Atlantic coast as far as Miami. He used conventional rod and reel, handline, fly tackle and even small harpoons, releasing most of the tarpon he caught regardless of how they were captured. *Book of the Tarpon*, which is still in print, is a must-read for those who want to know what fishing was like when Florida was still mostly a wilderness.

Mass-produced paddleboats slowly became more popular as fishing boats, enough so that in 1902, the Old Town Canoe Company incorporated, building its wood and canvas boats in Old Town, Maine. Some found their way to Florida. Old Town eventually became the world's largest builder of canoes.

During the late 1960s, Old Town pioneered the use of fiberglass in their boats. Shortly after, they introduced plastic and rotomolded boats into their line. This line now includes sit-inside kayaks, which are made using a design based on the ancient Inuit wood-and-sealskin boats, as well as a sit-on-top model. With the surge in the popularity of sit-on-tops and SUPs (stand-up paddleboats), canoes are less popular than they used to be, but Old Town and dozens of other companies still build and sell canoes.

In the meantime, other events were occurring elsewhere.

In 1971, Tim Niemier, a surfer and scuba diver in Malibu, California, who had just finished high school, modified an old surfboard to carry himself and his dive gear. Out of this humble beginning, the sit-on-top kayak was born. Niemier sold that first boat for three times his production cost and went into the kayak-building business. A few years later, he founded Ocean Kayak. This led to the huge popularity of the sit-on-top designs we see today, with dozens of manufacturers selling thousands of boats every year.

INTRODUCTION

Ocean Kayak produced the first sit-on-top kayaks.

Paddleboards have their devotees. They make fine fishing platforms.

Introduction

Like canoes and kayaks, the stand-up paddleboard (SUP) has a history that goes back thousands of years. Its modern form was developed in Hawaii in the 1900s, with its original use confined to surfing and related activities. Once the "modern" SUP reached California in the early 2000s, its popularity exponentially accelerated. By 2005, SUPs were used for racing, touring, river descents, yoga and fishing.

According to the website supconnect.com: "Then came the ultimate emancipation of SUP from its roots: fishing. Among the first was the Lane family down in San Diego, later a few people off Cabo San Lucas, but it wasn't till it reached Florida that SUP fishing became a certifiable chapter of SUP history. Corey and Magdalena Cooper, from Destin, Florida, launched a stand-up paddle company primarily dedicated to fishing, BOTE SUP." BOTE is one of many SUP companies now making vessels for fishermen.

Time, and paddle vessel evolution, marches on.

PART 1
ADVICE

1
BOATS

KAYAKS, CANOES AND PADDLEBOARDS

What boat should you use? I cannot answer that. People often ask me exactly that question. Why not ask me who you should marry?

Things to consider when looking for a vessel include your physical size, your fishing style, where you intend to use the vessel, whether you'd prefer to paddle or pedal, how well it paddles, whether you enjoy fishing alone, its cost, its weight, how you intend to carry it, its storage capacity, how comfortable it is, the aesthetics of its use and more. Everyone has different needs.

My fishing style when paddle fishing includes a lot of wading. I want a sit-on-top (SOT) kayak because I can easily get off of and on to the kayak.

One of my kayaks is an Ocean Kayak Drifter (no longer made) purchased in 2003. Designs have improved a lot since then. I still like that old boat for two reasons, though. First, my boats are carried atop my car. That one weighs fifty-five pounds, and I can lift it. Second, the boat is fairly wide. While searching for fish, I can stand up, allowing me a better view of the fishing area. Or, if it's an area where getting out of the boat is unlikely (deep water, soft mud bottom, etc.), this boat allows me to stand up and stretch.

This boat does not work well for camping. It lacks sufficient storage. I use a different boat for camping and am fortunate to have several different boats that I use for different purposes.

What about pedal kayaks? These boats have two main advantages for the serious angler. The muscles in your thighs are the largest muscle groups in your body. Pedal kayaks make use of these large muscles, which do nothing if you paddle. If you use your legs for propulsion, it frees your hands for casting. Paddling and casting don't compete with each other. Nick Lytle (navarrekayakfishing.com) has tried them all and thinks, at the time of this writing (spring 2019), that the Old Town Predator is the best one currently on the market.

Pedal boats are expensive (around $3,000 at the time of this writing) and heavy. Most are trailered or carried in the beds of pickup trucks. The drives don't work in shallow water, are subject to damage and are line catchers for fly anglers. Once again, only you can decide if one of these boats is right for you.

A sit-inside kayak.

What about sit-inside kayaks? Compared to a SOT, they are hard to get out of and into, so they don't fit my fishing style. This is strictly subjective, but I feel just slightly claustrophobic inside one. Most sit-inside kayaks paddle a lot more easily than SOTs, though. If touring is high on your list of reasons to get a paddleboat, you should certainly consider a sit-inside.

And, finally, stand-up paddleboards (SUPs), the type of paddle vessel I am least qualified to discuss, as I have used one only once. It was quite a windy day, and I tried to fly-fish. It was impossible, though, because the SUP moved so fast. I was like a sail on top of that thing. People who have brought them on my charters have had the same thing happen to them.

When the wind isn't blowing, the SUP works wonderfully. But it's windy a lot in Florida. For me, SUPs don't seem practical enough.

You, dear reader, are not me. You should get the boat that best serves the majority of your needs.

Any vessel you consider purchasing should get an extensive water test, preferably in snotty weather. If you can live with it in bad weather, you will love it the rest of the time. Only after this test should you part with your money. Borrow one, rent one, hire a guide—whatever you have to do to test it, do it. You can spend over $3,000 on a paddleboat these days; that's a lot of money to pay for an error. Make sure you like it before you buy!

Finally, for the remainder of this book, I will be using the terms *paddle vessel*, *paddleboat* or just *boat*. These terms all refer to all of the boat types discussed in this chapter.

2
ACCESSORIES

FLOTATION DEVICES, REQUIRED SAFETY GEAR AND OPTIONAL SAFETY GEAR

The U.S. Coast Guard requires paddlers to carry a few items of safety gear. For example, you need a sound-producing device, such as a simple plastic whistle.

If you might be out before sunrise or after sunset, you must have a single white light viewable from 360 degrees and visible at the distance of one mile. This should be atop a slender pole of some sort. If you are out at night, you are also required to carry three pyrotechnic devices (flares). These are good to have at any time.

You also must have a personal flotation device (PFD) for everyone on board. This can be a vest containing some type of foam flotation or an inflatable (by CO2 cartridge) vest. If you opt for the latter, it must be worn. When the PFD is not worn, it does not count as a PFD in the eyes of the law. Even if it's in your boat, if you're not wearing it, you're subject to getting a citation.

I like to paddle white water and, while doing so, have swum through some nasty places. Get a vest you like—one you will wear whenever conditions warrant it. Never be afraid or embarrassed to put it on. It can quite literally mean the difference between life and death.

Advice

Never be ashamed or embarrassed to wear a personal flotation device—it could save your life!

In Florida, you will see many paddlers with a long pole on the boat atop which is a small orange flag. If you paddle in fresh water, particularly where there are bulrushes and cattails, these flags warn powerboaters that you are there. Having one of these is a *really* good idea.

I charter out of a flats skiff and can testify from long experience that kayakers are hard to see from a speeding skiff when the light is bad. All three of my kayaks are bright yellow, because I do not want to get run over. My experience is that the fish do not care about the color of a paddleboat. Consider how you will be perceived by motorboaters when you buy your boat.

Every U.S. citizen should be required to take a first-aid/CPR class every other year. Since this will never happen, the responsible ones among us take it upon ourselves to do so.

You will be taking your boat into places where, if you get hurt, no one will be coming to help you. Learn first aid, and always carry a good first-aid kit. The day will come when you will need it.

PADDLES, ANCHORS, STAKE-OUT POLES AND PUSH POLES

Paddles! They're more fun to talk about than first aid. Regardless of the type of boat you use, you will usually propel it with a paddle. Over the life of that paddle, you will take tens of thousands of strokes with it.

A few years back, five of us took a paddle trip on the 160-mile long Indian River Lagoon. Before the trip, I invested in a Bending Branches canoe paddle, a fantastic piece of equipment—light, well-designed and well-built, aesthetically pleasing.

Another gentleman on the trip went to Walmart and bought an Indian Brand paddle—cheap, heavy, poorly designed and poorly made.

Yes, I spent more money. Yes, he finished the trip, too. But I will have my paddle—a treasured possession—until I die. One of my sons will get it. The Indian Brand paddle was falling apart at the end of the trip.

Get the best paddle you can afford. You will be less fatigued at the end of the day, you will enjoy paddling more and you will be the envy of your friends. That's fun, too. A new, decent-to-good paddle will run between $125 and $250.

Quality modern paddles generally have plastic or laminate blades and carbon-fiber shafts. Most kayak paddles come in two ferruled pieces. Paddles should fit both you and your boat. Any of the better paddle companies have sizing charts on their websites, as do all good paddle shops. Ask for help with getting your paddle fitted when you're buying it. You want a light, strong and well-fitted paddle.

If you can't afford the paddle of your dreams today, get the best one you can—for now—and save up for a better one. It's never a bad idea to have a spare.

One day, I fished in the no-motor zone of the Banana River Lagoon and was about four miles from the launch site. It was windy, so when fish appeared, I hopped off the kayak, tied the painter around my waist and started wade fishing. After an hour and a half, I had managed to scare all the fish away. As I was about to hop back on my kayak, I could not help but notice that my paddle was no longer on it. I did some quick math; the wind was blowing, and I had been off the boat for ninety minutes. The chance of finding the paddle was negligible. I did not have a spare.

Four miles of hard work that day taught me the hard way that two human hands make a terrible kayak paddle. If the wind was blowing the wrong way, I would probably still be there.

Advice

Bending Branches paddles are well-designed, well-made and aesthetically pleasing.

It's never a bad idea to have a spare paddle!

It's not a bad idea to have a paddle leash, either. Then, if your paddle does fall off your boat when you're busy or not paying attention, the leash prevents its loss.

A lot of Florida's saltwater paddle fishing happens in shallow water that is knee-deep or shallower. When it's too windy to effectively fish from the boat, you can just stake it out (or tie it to yourself), hop off and wade. My stake-out pole is a five-foot-long wooden closet rod whittled to a point at one end—very low-tech.

Being disabled, my friend Mike Conneen cannot use the wading technique. He has to stay in the boat. He has come up with a sophisticated way to control which direction he faces when he stakes out. He uses a custom-built stake-out pole made from the handle of a carbon-fiber canoe paddle and the bottom of a snow-ski pole tapered together with fiberglass and Bondo. It is attached to the boat by a leash and used with a kayak anchor trolley system

The anchor trolley is attached to the stake-out pole in this photo.

from Austin Kayak Company (www.austinkayak.com/products/22857/Yak-Gear-Deluxe-Anchor-Trolley-Kit.html).

If you want to fish in water deeper than about four feet, the stake-out pole doesn't work—you will need an anchor. Most kayak anchors weigh three or four pounds and fold up. Use the same trolley system I just mentioned.

Sometimes, you hook a large fish that requires you to immediately give chase. You have neither the time nor the spare hands to pull the anchor. Solve this problem before it happens by permanently attaching a float to your anchor line and attaching it to the trolley system with a stainless-steel carabiner. When Mr. Big comes along, unhook the clip and chase the fish until the (hopefully successful) conclusion. Then, paddle back to your float, clip back to the trolley and you're back in business.

A small folding anchor works for SUPs, too. Usually, it's attached to a D-ring somewhere on the board.

In my canoes, I use an eight-pound mushroom anchor. The float-and-clip idea works here, too, although I've never needed it while in a canoe.

This discussion of push poles mainly applies to canoes. SUP paddles are generally long enough to double as a push pole when one is in shallow water.

Advice

Poling a canoe when searching for flats fish makes a great deal of sense. *Courtesy of Roger Cook.*

You can certainly push a kayak with a pole, but when you see a fish, what do you do with the pole?

When using the canoe, the angler is in the bow, and the "guide" is in the stern. Of course, the guide can paddle, but he can see a lot better if he stands and uses a push pole. If the angler has good balance, he can stand, too. However, this is not for everyone.

My canoe push pole is a ferruled two-piece fiberglass pole made by Moonlighter Marine Products. Other companies make similar products. It's a fantastic piece of equipment. When I'm not using it, I can take it apart and put it inside the boat, where it remains (mostly) out of the way. When I need it, I just put it together, stand up and start poling.

AIDS TO NAVIGATION

I've never learned to use a sextant. It's a gaping hole in my education. OK, I've never needed to know how to use a sextant.

I can use a map and a compass, though. Nautical types call maps charts.

My first trip to Florida, back in 1980, included a six-day canoe trip through Everglades National Park. One thing that immediately struck me was how subtle the landmarks were there. Everything looked like everything else! While I had a chart and a compass, one of the big problems we had on that trip (and it remains a problem to this day with chart and compass use) was that much of the time, we did not really know where we were on the chart. Charts often have mistakes on them, and sometimes, you get mighty confused.

These days, you can buy a handheld GPS. But almost everyone already has a smartphone. If the phone has a signal, its map feature works as a GPS. In most places—but not everywhere!—in Florida, your phone will have a signal.

You'll have three problems with using your phone as a GPS. First, the phone does not like water, especially saltwater. You need a waterproof case for it.

Second, the battery will not last all day if you use the phone this way. You can buy a waterproof solar charger (they cost about sixty dollars) that will keep the phone running all day.

Finally, on many older phones, when the sun is bright, you can't see the screen. A large black shopping bag over your head helps, but this remains a problem without an elegant solution. If you have a solution, please send it to me via The History Press!

The author, surrounded by crocodiles, consults his chart. *Courtesy of Mike Conneen.*

Anyway, the combination of a good chart and a smartphone will keep you informed about your location in most places. If you are really worried about getting lost, look into a waterproof handheld GPS. I've been paddling in Florida since 1980 and have never needed one, though.

You need to find access to the water. This book lists many launch places but doesn't give directions. Why not? Everyone today has a computer or a smartphone. Siri gives better directions than I ever could!

Google Maps is an indispensable trip-planning tool at my house. If you don't already use this, you need to check it out.

The last part of this chapter concerns other resources you'll need. For this book to contain all the information you'd need to make a trip to any of the locations covered would require a much larger—and much more expensive—book. The appendix contains a short list of websites that have an enormous amount of information for paddlers.

SUN AND WEATHER PROTECTION

With the possible exception of insects, the sun will cause the unprotected paddler more distress than any other element of a trip.

I grew up in Massachusetts. When you were out in the sunshine, turning pink was a sign you needed to protect yourself.

If you see yourself turning pink in Florida, you are already way past well done. You need to protect yourself before you go into the sun here, because that sun is brutal and relentless.

The six elements of sun protection include:

- Plenty of drinking water
- A hat
- Good-quality sunglasses
- A buff (or other similar item of clothing)
- Sunblock with an SPF of at least 15, including a lip stick
- Lightweight long-sleeve shirt and pants

Younger, more virile readers may scoff at my precautions. All I can say is, I've lost friends to melanoma; carcinoma is no fun, either; and having the dermatologist use nitrogen to freeze precancerous lesions off your face and ears every year lacks any entertainment value. For that matter, your

Be prepared for rain when you paddle in Florida.

basic second-degree burn from sun exposure puts a damper on the rest of your trip.

Protect yourself now or pay the consequences later.

One of my favorite Lefty Kreh–isms is: "I have never understood why people think it won't rain while they're on vacation!"

I understand that while you're sitting in Cleveland planning your paddle trip to Florida, air temperatures in the fifties and falling rain are not part of your fantasy—but they could certainly be part of your reality. You need to be ready for that.

A good raincoat is an important part of weather protection. If you're here for the warmer months, that and a fleece shirt might be enough. For trips during the winter, you have to be prepared for cold weather.

Twice, I have been on trips in Everglades National Park when it got really cold; the worst time, the thermometer dipped down to twenty-seven degrees. Shivering, we wore everything we had, day and night, including our life vests. We used socks as mittens.

Fantasies are fine, but they won't keep you warm. Be ready for anything.

Advice

If you find yourself in Florida during the warmer parts of the year, you will witness Mother Nature displaying her awesome power in the form of a thunderstorm. You do not want to be on the water for this. You don't want to be near trees or anything else higher than its surroundings for this.

What you want to do is find some shelter that is lower than its surroundings. I like bushes, where you can patiently sit until the storm is totally gone before emerging. You'll find a raincoat quite a comfort in this situation.

WADING ACCESSORIES

Let me preface my remarks here by saying that my friend Rodney used to wade barefoot. OK, he was crazy. Still, it could be done.

That said, it's a bad idea. Broken glass, sharp shells, rusty metal, crabs, rays—need I say more? Your feet need protection.

During warm weather, I wet wade using flats booties made by Simms. Other companies make similar products.

In this waterproof backpack, I carry two DSLR cameras, my phone and my wallet.

When the water gets cold (for me, between Thanksgiving and Easter), I wear stockingfoot chest waders with cheap sneakers as the shoes. They don't have to be nice waders. They don't have to be stockingfoot waders.

You may think waders are overkill, but if you go wet wading in fifty-degree water, it's going to be a short fishing trip—and winter water temperatures often drop below fifty degrees.

ON KEEPING YOUR GEAR DRY

Plastic pollution is the scourge of modern civilization. However, when you want to keep something dry, a plastic-coated waterproof bag, a plastic bucket with a tight-fitting lid or a plastic cooler that locks shut keeps it dry.

Basically, your choices here are endless. The three items just mentioned come in an infinite array of sizes and shapes, and, of course, you can mix and match to suit your needs. If your stuff gets wet, it's only because of negligence.

3
PLANNING

LICENSES

Whether you fish in sweet or salt, you need a fishing license. See the appendix for details about how to get one.

TIMING YOUR TRIP

The best time to go fishing is whenever you can. However, you are not the only person planning a trip to Florida. The state hosts millions of visitors each year, and they all seem to come at the same time.

The busiest times of year on the water are spring break (March and April), Memorial Day weekend, Labor Day weekend, Thanksgiving weekend and between Christmas and New Year's Day. You can certainly visit during these times—you will not be alone, though. Where appropriate, make reservations as far ahead as possible.

Folks vacation during summer. Except for first thing in the morning, it's blazingly hot then. There's a lot of rain, and it can be buggy. Afternoon lightning lessens the appeal of paddle fishing, too.

Don't make a trip anywhere near the path of a recent hurricane.

All other things being equal, fishing is better during the week.

My favorite time of year is autumn, especially between Thanksgiving and Christmas. The light is fantastic, the leaves are changing color, the rivers run low, the weather is often incredible, the bugs are mostly gone and you just may have the place to yourself.

TRIP TYPES AND VEHICLE SECURITY

There are only two kinds of trips a paddler can take. One is the out-and-back. Most trips fit in this category. You launch your boat, park your vehicle, go out to fish and eventually come back to the spot where you started. Any number of people can do this. You can go solo, or you can bring a fleet.

On out-and-backs, use the tides (or current) and wind to your advantage when you can. Catch the tide out, catch the tide back. Go against the wind (or current, in a freshwater river) to start so you can paddle back with it when you're fatigued. Use your head instead of your back.

The other trip type is the through trip. You start at point A and paddle to point B. Unless you can somehow arrange for a ride (in some popular areas, companies will give paddlers a shuttle for a fee), through trips require a minimum of two vehicles and three paddlers. You unload your gear at point A, leaving one person there to guard the property while you and the third person drive both vehicles to point B, where you leave one vehicle. Then, you drive the other vehicle back to point A, where you start your trip. If you have multiple vehicles, leave as many at point B as possible.

When you make it to point B, the process is reversed. It's sad that you have to guard your gear, but if you don't guard your stuff, it may not be there when you come back.

Vehicle security will be an issue in some places. Never leave anything visible in your vehicle that a thief might steal (and thieves will steal anything). Lock anything you're not bringing in the trunk, or cover items with a towel or blanket or make them look unappetizing. That five dollars per boat launch fee may just be cheap insurance that no one messes with your car while you're gone.

BOATING ON SALTWATER

A piece of advice if you have not paddled much in saltwater: When you return to your vehicle, give your boat time to dry or towel it off before you load it. You don't want to drip salt all over your car—cars rust!

Rinse the boat with fresh water before loading if you can. Some paddlers carry jugs of water with them just for this purpose.

If you do get salt on your car, visit a self-service car wash (the type with the high-pressure wand) as soon as you can and wash the salt off the boat, the car, the tie-down lines, everything. Your car will thank you.

4
TACKLE

SPIN REELS, RODS, LINES AND LEADERS: ON LURE CHOICE AND COVERING THE WATER COLUMN

Again, keeping things simple is one of my biases.

When spin fishing in fresh water or inshore saltwater from a paddle craft, I carry one outfit. The standard outfit consists of a Shimano 3000 or 4000 reel, depending on the size of the fish in my fantasy. Other manufacturers make similar equipment. The 3000 reel will be loaded with fifteen-pound braid. The 4000 reel will have twenty-pound braid.

The rod will be light or medium light, six and a half or seven feet long.

The leader will be twenty-pound fluorocarbon, barring something unusual. We'll define unusual when we discuss different fish species. Twenty-pound fluorocarbon will not work for everything.

One small box carries all my terminal tackle. In that box, you'll find:

- An assortment of lead head jigs ranging from one-sixteenth to a half-ounce
- A selection of hooks, ranging in size from 3/0 to 5/0, for soft plastic baits
- Several three-inch DOA Shrimp (other manufacturers make similar products)
- Several DOA Bait Busters (other manufacturers make similar products)

- Several Johnson Minnows in gold and silver, one-eighth and one-quarter ounce
- Several five-inch Storm Chug Bugs with the treble hooks replaced with singles
- Several five-and-a-half-inch jerk baits

There may be a few worm rattles, barrel swivels, bait hooks and split shot rolling around in that box, with the rattles being the most important.

Somewhere in the boat, there will be one or more bags of three-inch shad tails. I like the DOA CALs and the RipTide Sardines. Other manufacturers make similar products. I might have a Cajun Thunder popping cork for desperate times. And there will be a spool of twenty-pound fluorocarbon leader material. That's everything.

With this small kit, I have baits that range in size from three to six inches long. Everything needed to cover the water column from top to bottom is included. I have baits that closely mimic what fish eat and baits designed to aggravate them into striking. My bases are covered. I can go anywhere in Florida and know that fish will bite what I have.

In the section about paddles, I said that a spare paddle could come in handy. This certainly applies to fishing rods. Keep rods inside the boat, out of the way.

You only need a small selection of lures. From top to bottom are a bag of Culprit plastic worms, DOA Shrimp, a quarter-ounce Johnson Minnow, a three-inch plastic shad and a three-inch plastic shad on a weedless jighead.

My friend Mike Conneen uses only four baits for most of his fishing. His go-to saltwater lure is the Vudu Shrimp. For a change of pace, he'll use a three-inch plastic shad that's brown with a chartreuse tail. In fresh water, he likes a small spinnerbait, with a plastic worm being the backup.

When spin fishing, I likewise use only four baits most of the time. My lure of preference is the three-inch plastic shad (I like the DOA CAL and the RipTide Sardine) on a 3/0 hook that has a one-sixteenth-ounce lead bead. Predator fish feed on this size of bait a lot! In deeper water, I'll put the same bait on a one-eighth-ounce weedless jighead. In fresh water, if I need a change, the reliable plastic worm is always in good taste. On the flyrod, I'll generally keep a popper but will switch to a streamer if conditions dictate that.

Understand, we have other lures (most with single hooks) and will use them as needed. My kit includes all the lures listed above. You don't need a ton of gear to catch fish, and having lots of stuff may be counterproductive if you spend more time changing baits than casting.

You should also have:

- A Dr. Slick hook file
- A stick of sunscreen for the lips (you can't have too much of this!)
- A package of Knot 2 Kinky leader wire—you never know when this might be needed
- A dehooker
- Dr. Slick chain-nosed pliers or equivalent
- A Dr. Slick line clipper

Leaders: The leader provides a (more or less) invisible connection between line and lure. For day-to-day use, a good-quality twenty-pound fluorocarbon is recommended. Now, let's take a longer look at leaders.

The thinner your leader is, the less visible it is to fish. Using a thin leader will lead to more bites. However, abrasion quickly weakens the leader. You will lose many of the fish that bite, along with your lure.

The thicker the leader is, the fewer fish you will break off. However, fish can see a thick leader, so you won't get as many bites.

The leader you choose is always a compromise between getting bites and losing fish. Some fish (bluefish, mackerel, sharks, tarpon and more) demand a thick leader, even wire. You will be best served by using the thinnest leader that will get the job done and accepting an occasional break-off as a cost of doing business.

Let us call the above the standard spin outfit. Remember what I said about finding what works for you.

Some situations call for an ultralight outfit. Fishing for sunfish, crappie and shad comes to mind. For this, the reel is a 1000 series with eight-pound braid mounted on a six-foot (more or less) ultralight rod. This ultralight outfit is much sportier when angling for smaller fish.

Some situations call for a heavy outfit. Big tarpon, cobia, mongo-sized crevalle and big sharks all require more beef. A 6000 series reel with thirty- to fifty-pound braid on a medium-heavy to heavy-duty rod will take care of these needs.

FLY REELS, RODS, LINES AND LEADERS: ON FLY CHOICE AND COVERING THE WATER COLUMN

Still working with the keep-it-simple theme, when fly fishing from a paddleboat (my preferred method), I carry a single rod, although there's often a spare inside the boat. A six-weight will work for most situations in fresh water and inshore. You can beef up if needed. A reel that holds a floating fly line and one hundred yards of twenty-pound Dacron fills out the outfit.

Leaders are constructed in two parts. The thirty-pound nylon butt is connected to the fly line with a loop-to-loop system. The twelve-, fifteen-, or twenty-pound tippet section, depending on season and species anticipated, is connected to the butt with another loop-to-loop system, complicated by the Bimini twist first tied into one end of the tippet section.

If you skipped the section on leader choice at the end of the conventional tackle section, you should read it. Everything that was said there applies to fly leaders, too.

Flies: Most fly patterns will catch fish. You need crab, shrimp and minnow imitations. You need flies that sink fast, flies that sink slowly and flies that float. You need flies that will attract fish and flies that will just aggravate them into striking. And, preferably, you need all this in a single small box and maybe a couple of ziplock bags.

My stuff gets carried in a Simms Dry Creek waist pack. In that pack, you will find:

- A one-quart ziplock bag containing a couple dozen synthetic minnow fly patterns, Puglisi-style flies, Polar Fibre Minnows and similar styles, in sizes from 4 to 2/0. Many have weedguards; some are tied as bendbacks

In that pack, you'll also find a small Plano box containing:

- 3 Dupre spoonflies
- A half dozen Merkin crabs, size 4, with weedguards
- Several Clouser Minnows in various colors and sizes (4-1) with weedguards
- Several black bunny leeches, size 2, with weedguards
- Several Borski-style Sliders, size 4, in various colors and weights, with weedguards
- Several unweighted bendback-style flies
- A few Rattle Rousers, size 4
- A selection of poppers and gurglers I can easily cast; these flies don't have to be huge

In addition, you should have:

- A couple of finger guards
- A Dr. Slick hook file
- A stick of sunscreen for the lips
- Small fluorocarbon leader wheels in twelve-, fifteen-, twenty-, and thirty-pound test
- A package of Knot 2 Kinky leader wire—you never know when this might be needed
- A dehooker
- Dr. Slick chain-nosed pliers (or equivalent)
- A Dr. Slick line nipper
- A small bag with a half-dozen small white shrimp flies for nighttime dock fishing—if you get a chance to do this, you will be ready

Let us call the above the standard fly outfit, but remember what we said about finding what works for you.

5
FISHING TECHNIQUES

SEARCHING FOR AND FINDING FISH INSHORE AND STALKING FISH

This first discussion concerns shallow-water inshore fishing only.

Like most other skills, you will find that learning to paddle fish effectively requires some paying of dues. You won't just hop in and start catching fish. You must learn how to use the boat to its best advantage. You'll have to learn how to handle the boat and how best to fish from it. Casting from a sitting position will require some practice if you haven't ever tried it before, especially if you're a fly caster.

And, you'll have to handle the boat and fish simultaneously. Both skills need to be completely second nature before you begin to have consistent success.

Let's look at what we might call a typical sequence to follow when paddle fishing inshore and hoping to sight fish. After launching the boat, paddle briskly to the area where you hope to find the fish. Once you've fished an area for a while, you will learn where these are.

Upon arrival, the paddling slows. You make your rod and reel ready to fish, whether it's a spin or fly outfit. The rod is ready to immediately deploy the bait, whatever it is. One of your jobs in all this is to maintain that state of readiness.

A fly angler casts to a tailing redfish while his friend controls the canoe.

You now hunt for fish, still moving, still covering water, but not brisk paddling that gets you where you're going. You want shallow water for this—less than a foot deep, if you're kayak fishing. You cannot see fish in water much deeper than this while sitting in a kayak. Shallow is good for redfish, big sea trout and snook.

Once you start seeing fish, you slow down even more. Then, when Lady Luck smiles on you, you see a fish that you want to stalk.

My experience has been that most folks want to get close enough to cast, then they flail away. This is a mistake. You want to slowly and silently put the boat into such a position that you can take a good, quality cast or a series of quality casts, one of which is likely to work. Once you get the boat into that position, you want to keep it there. Some people use a fancy anchoring system, some will stake out the boat. I prefer putting my leg over the side and holding it in place with my foot—another reason I like fishing in shallow water. If it's really shallow, the boat will literally be on the bottom, where it's not likely to move much.

Once the boat is in position and immobilized, then and only then does the casting commence. If you've picked the right position, and the fish hasn't done anything weird, one or two casts will usually be enough to get a response from the fish one way or another. In my own fishing, many times, there are only a few feet of fly line out of the rod, and I can often see the scales on the fish's back. Making a good cast in this situation is easy, and you can see everything the fish does in response to your presentation of the fly. There's no guesswork at all. It's deadly.

The above assumes several conditions can be met: shallow, fairly clean water; calm winds; and good visibility. It also assumes that you located some fish, which, fishing being fishing, won't always happen. Some trips just turn out to be a pleasant paddle through a beautiful area, and if you're past the beginner stage, you already know that such days come with the territory. You learn to enjoy them, and they make you appreciate the good days.

If you find any kind of structure—rock piles, oyster bars, potholes, channels, old pilings, etc.—it makes sense to work it hard. Fish like the security that structure provides.

What if it's windy? What if the water is too deep or too dirty? The basic approach remains the same, but you'll need variations on the theme.

In some boats, you can improve your ability to see by standing. Whether you can do this depends on your sense of balance as well as the design of the boat. If the ability to stand is important to you, you should get a boat that will allow for you to do it. Canoes and SUPs are easy to stand in/on.

A kayak angler has a choice of tailing fish to which to cast his fly.

This kayaker treats his boat like a stand-up paddleboard. *Photo by Cheryl Kumiski.*

Run the paddle vessel to those faraway fishing spots with a skiff.

Advice

My experience with standing while kayak fishing is that it doesn't work well. When you see a fish, you have to take your eyes off of it in order to put down the paddle (or pole) and pick up the fishing rod. In the meantime, the boat keeps moving and the fish keeps moving, and once you're ready, you have to visually find the fish again. If the fish realizes you're there, it becomes uncatchable.

If standing in the boat won't work for you, you can use the boat to get to the fishing spot and wade once you get there. If you work into the wind, you can tie the boat's painter around your waist and tow it behind you. If you go downwind, this won't work, since the boat will be in your way. You'll have to anchor or stake it out and come back and get it afterwards.

Another effective variation is to use a skiff to get your paddle vessel to a distant area. You can tow it or carry it in the skiff. Some guides specialize in transporting paddle fishermen to remote locations.

The fact is that most of us will figure out what we need to do for any given situation. There are more variations than we have space for here.

PRESENTING THE BAIT

Let's introduce the concept of the strike zone. The strike zone is an area around the fish where, if you present the lure well, you have a good chance that the fish will take it. This area is roughly shaped like a half a football, extending with the wide part at the mouth of the fish to the apex out in front of it. Understand that the strike zone constantly changes in size, going from nonexistent to huge and back again, and, occasionally, even goes behind the fish. Over the years, I've seen a few fish do an about-face to take a lure. It's rare, but it does happen. In order for a fisher to get a bite, the lure must be in the strike zone.

Hopefully, it's obvious that the longer the lure works through that strike zone, the more likely the fish is to take it. This brings us to presentation angles.

You could be positioned in front of the fish with it swimming at you. You could be behind the fish with it swimming away from you. You might be off to the side of the path of the fish, with your cast at a right angle to his line of movement, or you might be somewhere between one of these three locations. Where is the sweetest place to be?

The longer you can keep the lure in the strike zone, the more likely you are to get a bite. The place from which you can keep the lure in the strike zone for the longest amount of time is directly in the path of the fish with it coming right at you. You can adjust the speed of your retrieve to keep the lure a foot in front of it, or, if it's a school, in front of a lot of them. This is, by far, your best chance.

As you get closer to being at a right angle to the path of the fish, the sweetness of the angle declines. The lure stays in the strike zone for a shorter amount of time. A cruising redfish moves about three feet per second, so at a 90-degree angle, your lure might be in front of the fish for only a second or two.

In most areas, it's a mistake to cast past the fish and bring the lure to it. Prey seldom attacks the predator. Unless you've watched it happen, you'd be astonished at just how spooked a school of fish can be by a three-inch-long lure. If your cast is too long, it's best to wait for the fish to pass. Then, make another cast and try again.

Where the fish aren't pressured, that doesn't matter. In Louisiana, I was told, "As long as you get the lure within six inches of the fish's head, he'll hit it. The angle it's moving at doesn't matter." I had trouble believing this based on my experiences in Florida, but it was exactly right. However, Florida fish see more anglers than Louisiana fish. Your bait should not attack the fish.

As the fish gets past you, you're casting from behind it. You have almost no chance to get a bite. They see the line and leader land, and the lure always comes at them. This is a desperation tactic and only serves to spook the fish. It's much better to use the time to collect yourself for the next shot or to try to get back out in front of them.

One time, I had a gentleman from Atlanta on a charter in a canoe. He'd never caught a redfish and wanted one on fly, so I had taken him into the no-motor zone of the Banana River Lagoon. We didn't find a fish for a couple of hours, which was enough time for me to learn he was not much of a fly caster.

We finally found some fish, and there was a wad of them. They were huge. His longest cast required us to approach the fish too closely. Instead, I backed off. We got out of the canoe, hoping to wade into casting position. As we moved closer to them, I realized they were moving towards us.

"Cast, now," I said.

"I can't reach them," he responded.

"I know that, just cast!"

He made a thirty-five-foot cast that was thirty-five feet short. "Don't move that fly until I tell you to," I said. We waited, not breathing, as the fish got closer. Finally, they were clearly over the fly. "Just twitch it, now." He gave the

Advice

He caught that fish, a monster red pushing forty pounds, while wading.

fly two small hops, and the line came tight. The resulting battle was epic, and he caught that fish—a monster red pushing forty pounds—while wading in knee-deep water.

When casting to cruising fish, you want to lead them. In most places, if you hit them on the head, they respond in a negative fashion. Leading them is a must. When you lead a cruising fish or a group of cruising fish, the most important thing you can do at first is nothing. Wait until they are close to your lure before you move it.

At that point, all you need to do is make the lure look alive. A slight twitch is all it takes. Giant pulls—long, fast yanks—only spook them. Game fish don't have much trouble finding food most of the time. They usually don't need to feed aggressively. If you retrieve fast, they will usually disappoint you by ignoring your offering. Your goal is to make it as easy as possible for that fish to take your lure. Just make it look alive.

Sometimes, a redfish will follow your lure while trying to make up its mind whether or not to hit it. If it's a surface lure or an unweighted lure, just keep reeling. If you stop, they usually turn off. Frequently, though, they'll follow the lure to the boat and then see you, and then you're done.

If it's a weighted lure, stop reeling and let it dive to the bottom. Sometimes, the dive triggers the strike. If not, give it a short, sharp hop. If that doesn't work, you need to find a different fish or try a different lure. You can't catch them all.

Sometimes, you just know the cast and presentation should work, but it doesn't. I would give a lure maybe two chances like that, then it's time to change lures. Usually, when I change, it's to something smaller and darker. The only exception to that is when you have big fish chasing big baits, and they're ignoring your little three-inch lure. Then, it's time to take that giant thing with the 5/0 hook out of the box and heave it out there and see what they do.

Laid-up fish are the hardest fish to catch. They're not feeding, or they wouldn't be just lying on the bottom. I think when they're cruising or tailing, they're paying attention to their immediate surroundings, ignoring everything that's happening farther away. They don't see you coming, so they're relatively easy to catch.

When they're just lying there, they're looking for trouble. They see you coming, and a lot of times, when you make your cast, they leave before the lure even gets there.

The only way I have success with these fish is by wading into casting position. Use a small lure that kisses the water when it lands. Make it land a good distance away. If you're fishing a place with a soft mud bottom, wading is impossible. The fish will likewise be impossible. You can't approach them unobtrusively enough.

There's an old guide saying: "Never leave fish to find fish." Most of the time, this is a great adage. But, sometimes you have to realize that there are fish that just won't eat. Laid-up fish are often like that. Then, all you can do is try to find fish that are more active.

Fishermen get excited when they see fish, whether it's a single fish or a school of hundreds. When the time comes to make the cast, everything about it needs to be second nature. Things happen fast, and you need to be able to quickly respond to opportunities.

You need to be patient when that fish is there. One good cast is worth dozens of bad ones. Wait until you have a good angle, and be sure you can make the cast before you let it fly.

With some exceptions, Florida game fish usually don't feed aggressively. If you want a bite, you have to make it as easy as possible for them to eat. Get the lure in their face, and keep it there as long as you can. As soon as it's out of the strike zone, pick it up and put it back in there. If you do these things, you will have success. If you expect the fish to go out of its way to take your lure, you'll often be disappointed.

The single key to lure presentation is…make it easy for them to eat!

Advice

ON SWEAT EQUITY

Fish will never do quadratic equations, but they quickly learn what fishermen are—big trouble for them. The stupid ones are removed from the water. The smarter ones learn hook avoidance.

The farther you can get from popular spots, the easier the fish will be to catch. Part of the mission of this book is to share some of those places. The reader has to be willing to make the extra effort to get off the beaten path, though. Fortunately, paddle vessels are the ideal vehicle for doing this. Paddlers can easily access spots that never see motor vessels.

Don't be afraid to break a sweat. Your fishing will be better for it.

SEARCHING FOR AND FINDING FISH IN FRESH WATER

When fishing fresh water in Florida, you will generally cast at structure—weed beds, lily pads, stumps, pilings, fallen trees, drop-offs, points, pockets, rocks, that type of thing. In rivers, current seams are always good spots to try, as are holes at river bends, and the river's coves might be your best friend, especially during high water.

I still look for fish (there's no turning it off!). Rises, boils, flashes, shaking reeds, seeing them in the water—anything that makes me think a fish is there garners a cast or two or three.

The more thoroughly you work fishy-looking areas, the more likely you are to catch any fish that might be there. My friend Rodney says, "Ten percent of the time, the fish bite. Forty percent of the time, if you do everything just right, you'll get a few. The rest of the time, you're just fishing."

FIGHTING FISH FROM A PADDLE VESSEL

No special techniques are needed to play a three-pound sea trout or a six-pound redfish from a paddle vessel. Just reel it in.

If you stick a fish that's thirty pounds or more, though, technique becomes vitally important if you want to bring the beast to hand. We can't possibly cover every scenario here.

Before you go fishing, obtain a scale that reads up to about twenty pounds. Hook your line to it, hand it to a friend, move off forty feet or so, and try various rod angles to see which angle gets you the highest reading. The rod will usually be parallel to the ground (or water) or pointing slightly down. It's important to know how much pressure your tackle will take. It's important to have your drag set correctly.

Many different sources have advised setting the drag at 25 percent of the line's breaking strength. However, 10 percent is more realistic. Use a scale to learn what two pounds of drag feels like coming off the reel; when the scale is not available, you will still have a good idea about how to set the drag.

When a large fish realizes it's tethered, it generally zooms off at its highest speed. Your best strategy at this point is to try to stay close, keeping as much line on the reel as possible. When the fish exhausts that first rush of adrenaline, its pace will slow. This is when you go to work.

If the water is shallow, you can hop off the boat and fight it on foot. If the water is deep, turn your boat to increase its drag as much as possible. Place your feet over the side to accomplish the same thing. Make sure there

Fighting big fish from a paddle vessel requires good technique. This fish is a black drum.

are no large reptiles or sharks around before you do this, though—you don't want to get eaten.

While you pull on the fish, it rests. It finds the spot of least resistance and stays there unless you frequently change the angle of pull, so change the angle often! Make that fish work. When the fish is sufficiently rested, it will make another run. Let it. It will eventually tire and stop.

If the water is deep, the fish will get below you. You will have to lift it. Point the rod directly at it or as close to that as you can. As you lift (do not get the butt of the rod higher than parallel to the water), wag the rod back and forth. Changing the angle at which you pull will confuse the fish and bring it right up. It may go back down again, but after a few of those ups and downs, it will be exhausted and ready for boating.

If, during the process, the rod starts to come up too high, quickly reel it back down and continue to lift.

If you want to stop the drag from slipping while pulling on a fish, use your finger on the spool to increase the pressure. Then, if the fish surges, you can simply let go. If you tighten the reel's drag setting and the fish surges, it will break you off. Do not mess with the drag while playing a fish!

If the fish does not escape, you have to capture and unhook it—a moment of truth. I have an orange glove or a BogaGrip for these moments, but I prefer to use fingers. I'm not a fan of nets or gaffs because of the damage they cause to the fish. Some species of fish—especially large, toothy fish like king mackerel—pretty much demand a gaff, though.

HANDLING AND SUCCESSFULLY RELEASING FISH

Your fish has just completely exhausted itself fighting for its life. If you intend to kill and eat the fish, which is the *only* reason you should kill it, what happens now is moot. Photograph the daylights out of it. It's going to die anyway.

If you intend to release it, the best thing you can do to ensure post-release survival is to leave it in the water. Get your pictures with the fish in the water. When you're finished taking pictures, unhook the fish or cut the line, depending on what is safest and make sure it's strong enough to swim away with gusto before you release it.

The worst thing you can do to a big fish is put it on a gaff or a BogaGrip and hold it vertically, by its jaw, out of the water. Nothing in the fish's

Get your pictures with the fish in the water!

environment has prepared it for the resulting pull of gravity on its breathing mechanism or its guts. This is a horrible thing to do to a fish and ought to be outlawed. If you must put the fish on a BogaGrip, at least hold it horizontally.

You can find more information about this on the Florida Fish and Wildlife Conservation Commission website: myfwc.com/fishing/saltwater/recreational/fish-handling.

SEASONAL CONSIDERATIONS: HOW THE SEASONS AFFECT FISH AND FISHERMEN

Like human beings, all species of fish have preferred water temperatures. We can generalize that during the winter, the water is always colder than our fish prefer. We can further generalize that during the summer, it's always warmer than they like. If we assume that the water usually has enough dissolved oxygen for fish comfort, the fish move around in order to find sufficient food and the most comfortable water temperatures.

Your success as a fisherman is directly related to how well you can figure out where those places are.

During warm weather in the winter, you want to find places where the water is getting warmer. Late afternoons on sunny days are often best.

During the summer, you want to find places where the water is coolest. Early dawn or during the night is often best.

After hard winter cold fronts, when the water temperature is dropping, the only places you're likely to find fish are in deep canals and holes.

In mid-August, fishing in the middle of the afternoon, when the water temperature is in the mid-nineties, is often a fool's errand.

During the Florida summer, even the most diehard fishermen start wilting by 11:00 a.m. Add afternoon thunderstorms to the mix, and summer fishing becomes something that should be a dawn-patrol endeavor.

PADDLE FISHING SAFETY: USING YOUR HEAD TO STAY OUT OF TROUBLE

Seasonal considerations offer a great segue into fishing safety. Florida has sharks, stingrays, alligators, crocodiles, blazing sun, lightning, wind and horrible biting insects that carry equally horrible diseases. The two things you have to worry about the most as a paddle fisherman (besides not doing anything stupid yourself) are driving to and from the fishing spot and not getting run over by someone in a motorboat.

I can't say much about highway safety. While driving, people pay attention to things other than their driving, which makes driving dangerous. We all accept that when we get in our cars. If you want to drive legally, you must first pass a written test that demonstrates that you know the rules of driving. Then, you have to pass a road test to demonstrate that you can competently operate a motor vehicle. I have no doubt that our roads are safer because of these requirements.

In Florida, if you want to operate a motorized water vessel, all you have to do is get in and go. I do not trust people in motorboats to know what the rules are or how to operate their vessels. I avoid places where large numbers of motorboats are likely to be found. I suggest you do the same. One of the best things about fishing from paddle vessels is that it's so easy to avoid motorboats.

Let's discuss the other items from the first paragraph of this section.

- Sharks. They're here. They are not much of a threat. Do not hang fish from your boat with a stringer. While wading, do not hang fish from yourself with a stringer. Forget about stringers,

You could see a shark like this on any saltwater flat in Florida.

period! Caveat: You may well catch a shark. I have never found the value of a fishhook to be equal to the real risk of severe finger damage while trying to remove that hook from the shark. Cut it off as close to those choppers as you dare.
- Stingrays. These are a serious threat to waders. If you step on one, and there are lots of them, they will hurt you and result in a trip to the emergency room. When you wade, drag your feet rather than lifting them. I prefer wading in places where the bottom is clearly visible.
- Alligators and crocodiles. You'll only find crocs in Everglades National Park. Alligators are in every body of water in the state, including saltwater, and some of them weigh one thousand pounds or more. A few people get attacked by alligators every year. While the threat is small, it does exist. I know someone who survived a gator attack! Stay out of the water during low-light periods and avoid anywhere that you see one or more big alligators. If you see big gators around, keep your feet in your boat. If the gator looks like it's paying too much attention to you, go elsewhere. Do not hang fish

Advice

You could see an alligator like this anywhere it's wet!

from your boat with a stringer. While wading, do not hang fish from yourself with a stringer. Do not use a stringer, period!
- Sun. We have already addressed this.
- Lightning. Lightning strikes more people in Florida every year than in all the other states combined. A lot of those people are fishermen. If you see a storm building, even if the fishing is good, do some mental arithmetic. The equation is—more fish or my life? Find shelter as quickly as possible. I knew someone who did not survive a lightning strike.
- Wind. Wind produces waves. Waves swamp or capsize your boat. If the forecast says "small craft advisory," it's best to do something other than boating that day. It's hard to fish in a strong wind, anyway. If you must fish, try a protected area like a small creek.
- Biting insects. Bugs are part of the wilderness experience (not my favorite part, to be sure). I like a physical barrier—long pants, long-sleeved shirt, hat—that is two layers thick if necessary. DEET is awful stuff, but it works well. I spray it on my clothes rather than my skin. Head nets are available for extreme cases. You can also now buy special anti-bug clothing that is supposed to be quite effective.

If you find yourself in a small creek, there are two other things to watch out for.

- Poison ivy. This attractive plant loves Florida. A very common plant, it can grow as both vines and bushes. It can be on trees or low to the ground. In a small creek, you could find yourself wrapped up in it. Learn what it looks like, and learn to avoid it.
- Paper wasps. Although this is a low-probability problem, running into a hornet's nest is so unpleasant that it needs to be mentioned. Paper wasps sometimes build nests low over the water. Avoid them at all costs.

Above: Poison ivy leaves always have three leaflets.

Left: Running into a wasp nest would make for a bad day!

PART 2
THE FISH

According to the late Vic Dunaway's *Sport Fish of Florida*, the best Florida fish reference available for the nonscientific user (and highly recommended), there are 231 species of fish you can catch in Florida waters. We won't be able to cover all of them. The fish we'll cover are those most likely to be encountered by paddle fishermen.

1
FRESHWATER FISH

Our friends at the Florida Fish and Wildlife Conservation Commission have more information on this topic at myfwc.com/wildlifehabitats/profiles/freshwater.

LARGEMOUTH BASS: These don't need much introduction, as they're available all over the country. In our discussion of tackle, we did not mention plastic worms. If you're going bass fishing, have some plastic worms in a couple of different colors. One color should be red shad.

The redbreast sunfish is called the redbelly in most parts of Florida. They are beautiful little fish!

SUNFISH: Our standard outfits are too heavy for these fish. A two- to four-weight fly rod or a 1000-series spin reel on an ultralight rod are appropriate. Downsize your terminal tackle. It's hard to beat a small Beetle Spin when fishing for sunfish. The most common species available include bluegills, spotted sunfish (locally called stumpknockers), redbreast sunfish (called redbellies) and redear sunfish (shellcrackers). Others you might run into include warmouth, fliers and longears. All species of sunfish are delicious and make great targets when fishing with children.

Black crappie are common all around the state, delicious and fun to catch.

A modest channel cat from the Myakka River. It took a Vudu Shrimp.

CRAPPIE: The black crappie is distributed all across Florida. The best fishing for them happens during winter months. Use the same tackle you'd use for sunfish.

CATFISH: Several catfish species are found in Florida. The most popular with anglers is the channel cat. These can reach thirty pounds, so even though catfishing is primarily a bait-fishing proposition, it can be quite sporting. Popular catfish baits include cut fish, peeled shrimp and chicken livers.

STRIPED BASS: The St. Johns River is the southernmost extent of the range of these fine fish. You also find them in the St. Marys River and most rivers in the panhandle, where they sometimes reach impressive sizes. The state stocks striped bass–white bass hybrids, locally called sunshine bass. You fish for them the same way. These fish like colder water than most other Florida fish. Fishing for them is best during the winter months, when it's often slow for other species.

AMERICAN SHAD: These anadromous fish make their spawning runs in the St. Marys and St. Johns Rivers during January, February and March. Fly-fishers, in particular, get worked up about them. Our standard outfits are heavy for these fish, a large one of which weighs four pounds. Use the panfish outfit.

MISCELLANEOUS FISH: You may encounter chain pickerel and bowfin (locally called mudfish) while fishing here. You will certainly see gar, which can get fairly large. Carp have been introduced, generally for weed control, and get fairly large. While none of these species is generally targeted, all offer an interesting change of pace.

The Fish

The Mayan cichlid is one of many exotic species of fish you may encounter in the southern part of Florida.

EXOTIC SPECIES: In central and south Florida, you will encounter exotic species of fish that have been introduced, usually illegally. One the state brought in is the beautiful and popular peacock bass. These strike most types of lures that a largemouth bass would, although I don't think plastic worms are effective on peacocks. Tilapia and armored catfish (*Plecostomus*) are widespread. They eat algae and are hard to get on hook and line. Several species of cichlids are in south Florida. While small, they are quite aggressive and will strike small lures and flies. The oscar is established in south Florida. They are also aggressive and will strike small lures and flies. Snakeheads are also established in south Florida. Colorful and aggressive, they get to be about ten pounds.

2
SALTWATER FISH

Our friends at the Florida Fish and Wildlife Conservation Commission have more information on this topic at myfwc.com/wildlifehabitats/profiles/saltwater.

We find such a wealth of saltwater fish in Florida! Even covering just the most common ones likely to be encountered by paddlers, the list is long. Visit the Florida Fish and Wildlife Conservation Commission website (www.eregulations.com/florida/fishing/saltwater/) for information about size and bag limits and open seasons.

BLACK DRUM: If you see what you think are big redfish, but they ignore everything you try, you have found some black drum. Think of them as the redfish's big, dumb cousin. They range in size from little guys of a couple pounds to behemoths of fifty or sixty pounds—or more. You can catch them

The black drum is found statewide along the coast.

on artificials sometimes if you fish them slowly and right on the bottom. Thawed frozen shrimp, clams and pieces of crab work best for these fish.

BLUEFISH: Since they like colder water than most other Florida fish, more bluefish get caught during the cooler months. On the Atlantic side, they also tend to be small—three pounds or so. Gulf blues average a larger size. On both coasts, once in a while, a run of ten- to twelve-pound fish show up, creating quite the ruckus while it lasts. Bluefish have teeth like piranhas, so heavy fluorocarbon or wire leaders are needed when they are thick or running big. And these fish are aggressive—a fast retrieve often works when they ignore that slow retrieve recommended earlier. They strike almost every kind of lure.

FLOUNDER: A great prize for the table, you'll find flounder statewide. You'll have to work your lure close to the bottom to catch them, but flounder often lie in shallow water.

JACKS: Lots of types of jacks swim here. The most common is the awesome crevalle jack, *Caranx hippos*. Fast and aggressive, these fish love striking fast-moving surface lures and reach weights over thirty pounds. A thirty-pound jack can hurt you. Pound for pound, they are arguably the strongest fish in the water. Crevalle jacks taste terrible. Don't even try it.

LADYFISH: These sporty little fish have saved many a fisherman's day. When they hit twenty-four inches in length, they become quite respectable on light tackle. You do yourself a favor by never touching these or bringing them in the boat. The slime is like superglue, and they make a big mess. If you need chunk bait, ladyfish make a good one.

MACKERELS: The two species in Florida peninsula waters are the Spanish, which get up to six pounds or so, and the king, which often hit fifty pounds. Spanish mackerel will range inshore, but both species are generally encountered off the beaches. Both species have exceptionally sharp teeth. You can get by with thirty-pound fluorocarbon leaders for Spanish if you're willing to lose lures. Kings require

The teeth on a Spanish mackerel will cause severe cuts if you're careless!

wire leaders. Be careful when handling these fish! Like most oily fish, they are good eating if you care for them properly after the catch and eat them promptly. Neither type freezes well.

Pompano are a great prize—hard-fighting and delicious.

Redfish are popular inshore fish in Florida. This is a good one!

PERMIT AND POMPANO: Permit are some of Florida's great inshore trophies but are mostly found in the Keys. Pompano range inshore and along the beaches throughout the state. A big one is five pounds. They will take small jigs and weighted flies, particularly if they mimic mole crabs. Pompano flesh is highly regarded by gourmands.

RED DRUM (REDFISH): Statewide, these fish keep more guides working than any other fish. Redfish often feed in shallow water, offering an excellent sight-fishing target. They can reach weights of more than forty pounds. Redfish feed on smaller fish, shrimp and crabs, preferring to feed down. Your lure choice, whether using fly or conventional tackle, should reflect these preferences.

SEA TROUT: Florida waters host three species of sea trout, usually just referred to as trout. The one of greatest interest to anglers is the one that gets largest, the spotted sea trout. A good one will be twenty-five inches long or more and will weigh over five pounds. Trout and redfish often use the same areas. For dinner, trout like smaller fish and shrimp more than crabs and prefer to feed up, so they coexist well with the redfish.

SHARKS: Paddlers will see plenty of sharks in Florida waters. These range from little bonnetheads to giant tigers and hammerheads—scary and dangerous creatures. A few types will hit lures, especially popping plugs. If catching sharks is high on your list of things to do, your best bait is a small live or freshly dead fish, such as a pinfish, mullet or ladyfish.

The Fish

Sea trout like to feed up. These two were caught on fly rod poppers.

A small bonnethead shark, one of many species of sharks in Florida waters.

SHEEPSHEAD: Paddlers will see plenty of these striped fish as they search the shallows. Sheepshead seldom strike artificial lures. If you want to target the tasty sheepshead, live shrimp or fiddler crabs make the best bait.

SNAPPERS: This family of fish has lots of representatives in Florida. Paddlers are most likely to encounter the smaller inshore species, mangrove and lane snapper. Shrimp, small fish and pieces of small fish make the best baits for these tasty fish.

SNOOK: One of Florida's greatest inshore game fish is the snook. These fish behave somewhat like a largemouth bass with a bad attitude but can get much larger (over thirty pounds) than bass. Snook will strike all manner of artificials. Our standard twenty-pound fluorocarbon leader is too thin for snook, but if you use a heavier one, you will forfeit bites. Snook are tropical fish and love warm water. They are much more common in the southern part of the state.

The snook, an awesome game fish.

TARPON: Sometimes called the silver king, this is this writer's favorite fish. You can fish for little baby ones that are twelve inches long or giant fish that push two hundred pounds. The big ones can hurt you. Don't enter an engagement with one lightly. Tarpon strike all manner of flies and lures, and even big ones sometimes like surprisingly small baits. Like snook, tarpon like warm water (79 degrees is optimal). Unlike snook, tarpon are highly migratory. When the water temperatures are right, you can find them statewide.

This baby tarpon may live for thirty years and hit two hundred pounds.

WHITING: Most often found along beaches statewide, whiting is a small (up to three pounds) silver fish that is quite tasty. They'll take small jigs, especially if they're tipped with a piece of shrimp, and flies that are fished along the bottom. If the weather is calm and the beach is lightly populated (by people), you can sometimes sight-fish them along the shoreline.

PART 3
DESTINATIONS

This part of the book focuses on the best places for paddle fishing across the state—these areas are grouped together by chapter. I have used this—"($)"—to note where you can expect to pay a fee for use or service.

1
PANHANDLE WEST

Home to national seashores and forests, state parks, freshwater rivers, Pensacola, Choctawhatchee, St. Joseph, Apalachicola and Apalachee Bays, this area offers lots of choices for paddle fishers.

Because the Panhandle is so large and varies a lot, I split it into western and eastern sections. The western section (for our purposes) goes from Panama City west to the Alabama state line. The eastern section goes east from Port St. Joe to the St. Marks River. Let's look at the western section first.

If I were planning an inshore fishing trip to Florida, this is not where I would go. The three main towns in this stretch are Pensacola, Destin and Panama City. All are highly developed, very touristy and big draws for the spring-break crowd. It's not unusual to have jet-skiers buzz you while you're fishing, or to have some well-meaning but ignorant family in a rental boat motor up to you to ask how the fishing is.

That said, you can still find exciting fishing here. You just need to know where to look.

DESTIN

Destin, "the world's luckiest fishing village," has embraced beachside development in a big way. This reporter arrived to find throngs of beachgoers, jet skis, pontoon boats, sky-high condos, lots of kitchsy shops and restaurants and heavy traffic. He looked at the place as a giant tourist trap.

The view looking toward Destin.

Offshore fishing near Destin is reported to be some of the best in Florida. The red beach flags kept this kayaker from investigating. If you want to fish inshore here, you'd best start early. By noon, you're going to get trampled by the other boaters.

Chris Gatz generously offered to take me fishing. I met him at Liza Jackson Park in Fort Walton Beach. He had his posse with him. Four kayaks and a paddleboard were soon scouring Santa Rosa Sound looking for fish. We found some, too—sea trout, redfish and flounder. Chris told me other species are caught here all the time, including bluefish, ladyfish, crevalle jacks and Spanish mackerel.

We mostly looked around islands and along the south shoreline of the sound, but Chris does a lot of fishing around the numerous residential docks here. He says the night fishing around docks with lights is outstanding.

That night, I checked the weather forecast, which predicted winds blowing southeast at fifteen. I looked on Google Maps and found a boat ramp in Santa Rosa Beach on Hogtown Bayou on County Highway 383. The next morning, I launched my kayak there and started hunting.

The forecast proved accurate, so I stayed on the lee shoreline. The water was clear. The bottom was covered with seagrass that had potholes of all

Fishing near the Intracoastal Waterway near Fort Walton Beach.

sizes in it. I worked it hard with a jig, with the only reward being a single, modest-sized sea trout.

When the sun got up high enough, I stood up and started looking for fish. There were lots of them, both sea trout and redfish. I could not get a bite, though.

Several other boats showed up, so I worked my way back to the boat ramp. There, I met a fly-fisherman, a member of the fly fishing club in Destin. He told me where I fished was the most heavily fished part of the bayou, and fishing was better on the north side.

Next time, I'll know.

The next morning, I wanted to try fishing in Basin Bayou. Lightning, heavy winds and heavy rain caused me to change that plan. However, the weather cleared around noon, so I drove along State Highway 20 until I found the launch ramp at Nick's Seafood in Niceville (which is an awesome name for a town). Then, I launched the kayak and paddled into the bayou.

In the southwest corner of the bayou, there are a handful of houses. Other than that, the pond's shoreline is completely wild and filled with cypress, red maple, slash pine—the usual suspects. Ospreys and bald eagles patrol the skies, while barn owls hoot from the woods.

This bass is a resident of Basin Bayou.

The water is tannin-stained but clear. Thick beds of *Vallisneria* carpet the bottom, and rafts of water lilies hide in tucked away corners. On the north side of the bayou, an explorable stream enters.

Fish species include freshwater types like largemouth bass, sunfish and chain pickerel and saltwater types like redfish and sea trout that come in from Choctawhatchee Bay. Once your fishing is over, you can enjoy some oysters and a cold beverage at Nick's (www.nicksseafoodrestaurant.com). Basin Bayou is beautiful and not to be missed.

Nuts and Bolts

Destin Visitors Center:
   www.emeraldcoastfl.com
Boat Ramps: There are a lot of them here.
   fishingdestinguide.com/BOATRAMPS.html

Destin boasts a fishing museum, unsurprisingly called the Destin Fishing Museum. While they're big on photos of large dead fish, they have some worthwhile displays. It's a great place to visit when lightning flashes and rain pours. www.destinhistoryandfishingmuseum.org

Rentals

Destin Kayak Rentals:
850.974.1787, destinkayakrentals.com

PANAMA CITY

Having been to Panama City several times, I planned my traverse of town to occur at about 7:00 a.m. on a Sunday morning, thereby missing all the traffic that would otherwise be encountered. I wanted no part of spring break central. The place is a GIANT tourist trap.

If you must:

Nuts and Bolts

Boat Ramps:
fishingdestinguide.com/boatramps-saltwater-PANAMACITY.html

Rentals

Mr. Bill's Kayaks:
850.767.9913, mrbillskayaks.com

PENSACOLA AREA

During a recent visit to Pensacola, we camped for a couple of nights at Fort Pickens campground in Gulf Islands National Seashore. The campground lies near the western tip of Santa Rosa Island, a long barrier island with a spectacular beach. There are kayak access points on the Santa Rosa Sound side of the national seashore. A kayak dolly will come in handy—it's one hundred or more yards from the parking lot to water.

Santa Rosa Sound boasts clear water, scattered seagrass beds and many species of fish, including sea trout, redfish, bluefish, Spanish mackerel, ladyfish, crevalle jacks, pompano and more. The best fishing here happens between mid-April and mid-October. The jacks, in particular, reach hurt-you sizes—twenty pounds or more. Never take a twenty-pound jack lightly!

Most of the north side of Santa Rosa Sound is private property, but the national seashore visitor center in Gulf Breeze has access to the water for paddle fishers. The fish here see lots of anglers, and they're tough, but there are plenty of them.

In many places, on the bottom, you will see lots of tiny mounds that look like microvolcanoes. Each mound is home to a ghost shrimp, a favorite food of sea trout and redfish. These critters are light in color, nearly transparent. Your choice of lure should take this into account.

FORT PICKENS

In an ironic twist, the Army Corps of Engineers used hundreds of slaves to build Fort Pickens. Building the fort was a miserable endeavor—over 20 million bricks had to be laid by hand amidst heat, humidity and insects to create the five-sided fort with its four-foot thick walls; the five-year project was completed in 1834.

While the fort was built to defend Pensacola Harbor against foreign invasion, the only combat the fort ever engaged in occurred during the Civil War. During the war, the slaves who survived the building of the fort were freed.

The ruins of Fort Pickens; an explosion blew out the wall.

At one point during the war, Confederate forces tried to wrest control of the fort from its Union garrison during an assault called the Battle of Santa Rosa Island. The rebels failed. Fort Pickens remained in Union hands throughout the war.

After the war, the fort was used to house Apache prisoners, including Geronimo, who had been defeated and transported from the American Southwest. They became something of a tourist attraction while they were here.

PENSACOLA PROPER

We stopped at the tourist information center to get information about the town. Behind the building is a seawall. While we were watching, a school of crevalle came in and terrorized the mullet that were there. A local fisherman told us that this happens regularly during the warmer months of the year.

Another ongoing opportunity for fishing occurs at night, when big redfish are taken around the bridge. Given the currents, water depth and darkness, I would not do this type of fishing from a paddle vessel. Some local folks do it all the time.

SURROUNDING AREAS

We met Nick Lytle (navarrekayakfishing.com) at a convenience store at zero-dark-thirty. We were a trifle bleary-eyed, but as the sky lightened, we followed him for about thirty minutes before he stopped next to East Bay. It was cold enough that steam came off the water.

We launched our vessels, paddling and casting as we worked our way down the shoreline. Broken, golden and blessedly warm sunlight filtered through the trees. The fish were nowhere to be seen.

Nick said, "When the sun gets above the trees, the bait, and the fish, will start to move." Never doubt the guide. We waited patiently, trying to keep ourselves in the sunlight. Some of us were shivering.

Sol reached a point approved of by Nick. We went hunting. I flushed one redfish, then another. A large trout swam past, leering at me, refusing to bite.

Nick hooked and caught a junior-sized redfish. A pair of tanks swam in to range. I blew the cast.

The rest of the morning, we had continuous shots at redfish of all sizes in all kinds of numbers and configurations, from singles to groups of fifty or so. Nick caught three. I hooked and lost two. We didn't work it hard, and when motorboats started showing up, we headed back to the launch.

Maybe we should have caught more fish, but it had been a spectacular morning.

East Bay and Escambia Bay are both large. Both have many feeder streams and bayous. My visit was not long enough to begin to adequately explore either.

One evening, Jim Tedesco and I checked the weather forecast: southeast at fifteen, a recurring theme during my trip. Using Google Maps, we found the Escribano Wildlife Management Area on the east side of East Bay. It appeared that if we went into Catfish Basin, we would be protected from the wind.

Kayak guide Nick Lytle with a nice redfish.

This sea trout was the result of detective work, hard paddling and some good luck.

Upon our arrival the next morning, the wind still behaved. We launched our kayaks and headed toward the basin, casting as we went. The fish did not disturb us as we paddled.

Once we got into the basin, though, we started hitting fish. They were all sea trout. None could be termed trophies, and many weren't even legal. But we had some action.

Unfortunately, by this time, the wind was no longer behaving. The breeze precluded our exploration of the basin. Getting back to the car was hard, wet work. Cold, choppy waves slopped over the bow of the kayaks as we paddled back to the launch site.

But we had picked a spot off the map, found it, found some fish and had a good time. It was all you could hope for on a windy day somewhere you had never been.

OFF THE BEACHES

Along the public beaches in the western Panhandle, the lifeguards fly flags to let you know how rough the surf is. The flags are red, yellow and green, just like traffic lights. The entire time I was in the Panhandle, red flags flew. I did not get to fish off the beach.

Nick Lytle tells me it's great fishing. Little tunny, Spanish and king mackerel, tarpon, red snapper, cobia and even sailfish are all possibilities.

If you have not kayak fished off the beach before, your first time should be with a guide. Summer weather is the most stable. I have every confidence in Nick Lytle. If I were going to Pensacola, I would certainly contact him first.

FRESHWATER OPTIONS

BLACKWATER RIVER

Snow-white sandbars contrast with clear, shallow waters lined by magnolias, cedars and a variety of wildflowers. On the Blackwater River, you will notice a similarity of the snow-white sandbars to the snow-white beaches of the Emerald Coast—it's the quartz sand! Over the ages, sand from the Blackwater River took a trek through pools, swamps and the bays until it reached the Gulf of Mexico. There, it washed up to form some of the world's most beautiful beaches—those of Santa Rosa Island.

The average depth of the Blackwater River is two to three feet and its average speed is five miles per hour, with clear, clean water and a white-sand bottom. Fish species include bass, sunfish and catfish. Most of the way, you paddle through a state forest.

Blackwater Canoe Rental rents paddle vessels and offers a shuttle service with two- and six-hour (paddling—not fishing—time) trip options. Camping is an option. Visit during the week! www.blackwatercanoe.com.

COLDWATER CREEK

Like the nearby Blackwater River, Coldwater Creek flows through Blackwater State Forest, and like on Blackwater River, a livery provides rentals and shuttle service. Adventures Unlimited offers four-, seven-, eleven- and fifteen-mile trips, and yes, camping is available. You'll find the same fish species in both creeks. adventuresunlimited.com/river-adventures

Destinations

ESCAMBIA RIVER

Sadly, I did not fish here. This water is home to eighty or so different fish species.
myfwc.com/fishing/freshwater/sites-forecast/nw/escambia-river

My friends at the Florida Fish and Wildlife Conservation Commission tell me that the Wacissa River and the Chipola River are also worth visiting. I did not get to fish either of them, unfortunately. So many fish, so little time!

Nuts and Bolts

Gulf Islands National Seashore is a must-visit if you're near here:
www.nps.gov/guis/index.htm
Pensacola Visitors Center:
www.visitpensacola.com

Pensacola has a rich and colorful history of which it is justly proud. You'll find several fascinating museums. The folks at the visitors center will put you in touch with it.
Pensacola area boat ramps: There are a lot of them.
fishingdestinguide.com/boatramps-saltwater-PENSACOLA.html

Rentals

Littleheads Kayak Rentals, Perdido Key:
251.284.5107, www.littleheadskayakrentals.com

PENSACOLA HISTORY

Pensacola's fine deep-water bay has been coveted by every seafaring nation that has seen it. When the Spaniards arrived here in eleven ships in 1559, several tribes of indigenous people were already here. The Europeans ignored them and started setting up a colony. One thousand people brought livestock, tools and the materials and supplies they would need to build a town.

A month later, a powerful hurricane struck, sinking most of the ships, which were being used as warehouses. The survivors were eventually evacuated. The Spanish did not return until 1698, at which time they built several small forts here in an attempt to slow French incursion on the Gulf coast.

Since the first Spanish colonization attempt, the Spanish, French, British, United States, Confederate and again United States flags have flown over Pensacola. The old part of town was laid out by the British.

Florida was the third state to secede from the Union at the start of the Civil War. While Fort Pickens remained Union, Pensacola did not. In 1862, Pensacola was retaken by Union forces, and most of the town was burned.

Until the late twentieth century, this part of Florida was a sleepy backwater. Since then, there has been dramatic growth in the beach-based tourism industry and rapid development of previously pristine wilderness beaches. The establishment of several state and national parks has preserved some of those beaches from the encroaching high-rises and condos.

2
PANHANDLE EAST

The region from Port St. Joe to St. Marks is called the Forgotten Coast. St. Joe Bay has a reputation for being the healthiest bay in the state. The lifestyle here is laid-back, the fishing superb. For a fishing vacation, this is the place to go.

ALLIGATOR POINT TO ST. MARKS

This is a wonderful area to fish—great fish habitat, lightly populated by people, fantastic places to paddle fish.

It's worth exploring inside Alligator Harbor, especially out by the point itself, which is an awesome place. Northeast of Alligator Harbor, you'll find Bald Point State Park, which makes a great base camp for exploring this whole area. Ocklockonee Bay and Apalachee Bay meet at Bald Point, creating a maze of sand and oyster bars—a huge fish magnet.

If you head east from Bald Point along the south side of Ocklockonee Bay, you'll find a mostly undeveloped shoreline with numerous feeder creeks emptying into the bay—obvious spots to fish. You can explore and fish all the way back to the mouths of the Ocklockonee and Sopchoppy Rivers.

Driving north along State Highway 98 over the Ocklockonee Bay bridge, you'll find yourself in the village of Panacea. You can access the north side of the Ocklockonee/Apalachee Bay confluence at Ocklockonee Point from

A redfish caught near Bald Point.

Mash Sands Beach. From here to St. Marks, the shoreline is all salt marsh, with numerous oyster bars and lots of islands. If you can get your vessel in the water, you are in a place where you can catch fish.

Finally, much of the land here is included in the St. Marks National Wildlife Refuge (www.fws.gov/refuge/St_Marks/visit/visitor_activities.html). There's fresh- and saltwater fishing to be done. Often overlooked, this is a fantastic area to explore.

Nuts and Bolts

First, any visitor to this area during the second Saturday of April must plan on attending the Sopchoppy Worm Grunting Festival. No, I am not making this up. Real human beings grunt worms out of the ground as a contest, and a Worm Grunting King (apparently, the worm grunters are all male) is crowned. Food and art vendors, live music and more make this a unique celebration.

Wakulla County Visitors Bureau:
   www.visitwakulla.com

Launch sites:
   www.saltchef.com/catch_fish/FL/Wakulla/boat_ramps.html

DESTINATIONS

Rentals

The Wilderness Way, Crawfordville:
  850.877.7200, thewildernessway.net
Wacissa River Canoe and Kayak Rentals:
  850.997.5023, www.wacissarivercanoerentals.com

APALACHICOLA BAY AND ST. GEORGE SOUND

Redfish, oysters and tarpon are Apalachicola Bay's main draws. No matter what style of saltwater fishing you prefer, you can find it here. Tidal rivers, salt marshes, oyster bars, flats, beaches, the bay itself and the open Gulf all beckon paddlers.

Anywhere along U.S. Highway 98 that you can park, you can launch a paddle vessel. As long as you are not in the vicinity of Apalachicola River, the water is usually clear. Good areas to try include the stretch along Highway 98 between Yent's Bayou and Carabelle Beach and on St. George Island in St. George Island State Park.

Releasing a St. George Sound sea trout.

Spanish mackerel are found all along the Panhandle coast.

Weather has some say as to where you fish, as does the season. Points, passes and river mouths are all good places to try. Oyster bars are fish magnets if water flows over or around them. Diving birds create an obvious opportunity.

During the summer months, tarpon come into the bay after menhaden. These tarpon are big fish, not to be taken lightly.

During the winter, when not much else happens, striped bass in the Apalachicola River provide a solid fishery. These fish run from about three to ten pounds.

Last time I was there, I went fishing in St. George Island State Park ($). Two minutes after launching the kayak, I caught a solid trout on a jig. I put the spin rod away and started blind casting with a Clouser Minnow. The fish were not suicidal, but they came steadily—good ones, trout to four pounds, reds to eight. Got a fine Spanish mackerel and had another cut me off. It was cloudy, so I could not sight fish, but it didn't matter.

When I tired of casting the fly in the wind, I switched back to spin tackle and a weedless jig. If anything, it was even more effective. And I did not see another fisherman the entire day.

DESTINATIONS

FRESHWATER FISHING

You'll find lots of public land here with many fishable waters. I hate to report that I haven't fished any of it, but there you go. Florida is big, and one has only so much time. If you'd like to explore it, start your planning with the Lower Apalachicola River Corridor Road Map to Recreation: apalachicolareserve.com/documents/RoadMap.pdf.

Further planning help can be secured at the following websites:

Apalachicola National Forest:
  www.fs.usda.gov/activity/apalachicola/recreation/fishing
Apalachicola River Water Management Area:
  www.nwfwater.com/Lands/Recreation/East-Region-Recreation
Apalachicola River Wildlife and Environmental Area Paddling Trail:
  myfwc.com/viewing/recreation/wmas/lead/apalachicola-river/things-to-do/paddling
Box R Wildlife Management Area:
  myfwc.com/viewing/recreation/wmas/lead/box-r/things-to-do
Tate's Hell State Forest:
  www.freshfromflorida.com/Divisions-Offices/Florida-Forest-Service/Our-Forests/State-Forests/Tate-s-Hell-State-Forest

Rentals

Island Outfitters:
  850.927.2604, www.sgislandjourneys.com

History

When cotton was king, Apalachicola had a corner on the market. The fiber was shipped down the Apalachicola River in bales, from the interior, by steamboat. Ships from all over the world came to Apalachicola for cotton and lumber. Forty-three cotton warehouses once lined the riverfront. Revenues from the lucrative cotton trade financed many fine mansions, several of which still stand despite a disastrous fire that occurred around 1900.

Today, the Raney House is operated as a museum by the town of Apalachicola. The Orman House is a Florida state park and museum. Other historic homes are in use as inns or guesthouses.

The first ice machine was developed in Apalachicola by Dr. John Gorrie, who received a patent for his invention in 1851. His reasons for wanting this machine were mostly altruistic; as a physician, he was trying to lower the temperatures of patients affected by yellow fever.

Gorrie's invention became the basis for modern refrigeration and air conditioning. Today, Dr. Gorrie's house is also a Florida state park and museum.

ST. JOE BAY

This bay offers protection from all but north winds, beautiful clear water, great wading in most places and a wide variety of saltwater fish, including sea trout, redfish, Spanish mackerel, flounder, pompano, bluefish, kingfish, tarpon, cobia and more. You can stay in luxury in one of the many rental waterfront condos or camp at St. Joseph Peninsula State Park, a great facility. The state park has cabins for rent, too, although I've never stayed in them.

The bottom of the east side of the bay has (generally) a stair-step arrangement—a shallow flat, largely exposed at low tide, drops off to a narrow, deeper flat, which drops off again to a still deeper area. The deeper parts of the bay have over thirty feet of water. Big fish come in here.

The south and west sides of the bay have extensive shallow flats covered with thick seagrass beds. They're interspersed with sand holes and sandbars, eventually dropping into the depths. It's gorgeous and paradisiacal.

From the kayak launch in the state park to the tip of the St. Joseph Peninsula lie seven miles of limpid water, seagrass beds interlaced with white-sand potholes. For those who sight fish, it borders on heavenly.

In the event you find yourself blown out here because of a north wind, you can simply take your vessel to nearby Apalachicola Bay—hardly a drop-off as far as quality of fishing is concerned.

You can launch a paddle vessel in St. Joe Bay at the city boat ramp ($); at Parnell's Marina, also on the east side of the bay ($); at Salinas Park, in the southeast corner of the bay (a kayak dolly comes in handy here; it's a long carry); at the Lighthouse Bayou paddle launch in the southwest corner of the bay; and in the state park ($).

Fat flounder are just one kind of fish you'll find in St. Joe Bay.

The kayak launch spot at St. Joseph Peninsula State Park.

Nuts and Bolts

Gulf County Visitors Center:
  www.visitgulf.com
Kayak launches can be found all over Gulf County:
  www.visitgulf.com/adventure-map

The website for the state park is www.floridastateparks.org/park/St-Joseph. If you hope to camp, reservations are recommended.

Rentals

Happy Ours Kayak & Bike Outpost:
  850.229.1991, www.happyourskayak.com
Scallop Cove General Store:
  850.227.1573, scallopcove.com
Barefoot Kayak & Paddle Board Rentals, Mexico Beach:
  850.899.0359, www.barefootkayaking.com
Presnell's Marina:
  850.229.9229, presnells.com

History

In 1835, speculators founded the town of St. Joseph, hoping to compete for the lucrative trade in cotton and lumber that Apalachicola enjoyed. A railroad was pushed through the (then) wilderness to the new town. St. Joseph Bay was an excellent harbor, but there were no feeder rivers that could carry shipping from the interior. The railroad was a commercial necessity. The boom town grew rapidly, reaching a population of about 12,000 in just a few years.

St. Joseph had enough money and influence that Florida's first Constitutional Convention occurred here. Yes, Florida's constitution was drafted in a town that no longer exists.

By 1840, the town fathers could see that St. Joseph could not commercially compete with Apalachicola. So, they changed the format, opting to become what would now be known as a sinful resort town. Gambling, whiskey and loose women became the town's calling cards. I can imagine the Tallahassee politicians of the day good-naturedly elbowing each other in the ribs and whispering, "What happens in St. Joseph stays in St. Joseph."

In 1841, the format changed again, although this time, it was an accident. A ship coming in from the Antilles brought yellow fever. The disease quickly became a raging epidemic that killed over half the town's population. Those who survived abandoned the town.

Two years later, the storm surge from a major hurricane demolished what was left of St. Joseph.

In the early twentieth century, a new town was founded near the site of St. Joseph. In 1936, Port St. Joe became a paper town, with a paper mill providing most of the jobs for residents until 1996, when it was shut down and dismantled.

Today, the town primarily survives on tourism.

INDIAN PASS

If you're motoring along Highway 98, you could drive past Indian Pass and never realize it. For a paddle fisher, this would be a mistake.

From the boat ramp on the west side of the pass, you can paddle a half mile to St. Vincent's Island, a national wildlife refuge, for some exploration and both fresh- and saltwater fishing or paddle north into Indian Lagoon and fish for flounder, sea trout, redfish and more around the plentiful oyster bars and islands there. This is one of the Panhandle's least-fished areas and is worth a day or two of exploration.

A private campground, Indian Pass Campground (www.indianpasscamp.com), is located here and makes for a convenient base camp.

3
NORTH FLORIDA

BIG BEND PADDLING TRAIL

Take day trips or camp for weeks along one of the least-developed coastlines in Florida. Miles of salt marsh, oyster reefs and seagrass beds support a variety of saltwater fish.

A BIG BEND PADDLE ADVENTURE

Rudely, a harsh wind punched Mike Conneen and me right in the face. We rounded the point, paddled into the Gulf and BOOM! The four miles we had to go straight into this stiff breeze seemed a daunting task. Mother Nature chose to test our mettle from the beginning. No laid-back first day here!

    The Florida Fish and Wildlife Conservation Commission (FWC) maintains a series of campsites along a 105-mile length of the Gulf Coast in the Big Bend region from the Aucilla River to the town of Suwannee. You'll find these sites, specifically reserved for paddlers who wish to make a trip like the one we were on, placed about an easy day's paddle apart—an easy day's paddle apart, that is, assuming a fifteen-plus-knot wind isn't slamming you in the face. That wind significantly increases travel time and the necessary sweat equity.

Paddling hard on the Gulf of Mexico along the Big Bend Paddling Trail.

This stretch of coast between the Aucilla River and the Suwannee River lacks much dry ground. What's there is already occupied. The FWC deserves commendation for maintaining these campsites. Some are spectacular. Others are marginal. If you go, be prepared for bugs and mud. Bring something on which to sit. And don't forget that you need to have a permit to use them (see Nuts and Bolts at the end of this section).

The FWC also makes the *Big Bend Saltwater Paddling Trail Guidebook* available online for fifteen dollars. It's highly recommended to anyone seriously considering making this trip. Again, see Nuts and Bolts for ordering information.

We barely passed Mother Nature's first test and reached the mouth of the Econfina River late in the afternoon. Luck would have the tide dropping, of course, so in our subdued state, we had to paddle another three miles against the current to reach the campsite. There was mud, and there were bugs, and we turned in as soon as darkness fell. I listened to the sound of critters and insects and the now-soothing sound of the wind through the treetops and wondered what I'd gotten into.

The draw for a trip like this is not to fight the wind all day, of course. It's fish, birds, falling stars and planets and constellations, crabs and turtles and seagrass. It's getting away from traffic and telephones and email and clocks.

Look at Florida's Big Bend on Google Maps, and you'll see that getting away from modern life might be easy to do here. Countless creeks dissect the coastline, which consists of scores of miles of mostly untouched spartina marshes, oyster mounds and lush grass flats. You find water so clear it's crisp. In that water swim the fish that drew Mike and me—redfish and sea trout, flounder and bluefish.

Day two required eleven miles of paddling—three to the Gulf and eight more to our next campsite. The wind had not changed in direction or intensity. Our technique was simple: paddle to a point, rest. Paddle to the next point, rest. Repeat as necessary. We crawled along the coast this way as the earth hurtled through space. When we finally reached Rock Island (one of the spectacular campsites), it was late in the afternoon. We were beat.

You'll need a permit to camp on Rock Island, a beautiful place.

Sunset promised to be glorious. During dinner, I said to Mike, "After today I wouldn't ask you to do it, but you fishing against the sunset would make a great photo." As soon as we finished eating, Mike jumped into his kayak and paddled toward the sunset. I got the photo. Mike got a couple fistfuls of bluefish—our best fish of the trip to this point.

That night, I listened to the night sounds. There were no bugs, no critters—only waves and wind.

FWC campsite permits allow only one night at each site. When we got up the next morning, the wind remained unchanged. Whitecaps covered the Gulf. Our next campsite, which was ten miles away, required us paddling straight into wind and waves.

Mike said, "I feel like I went twelve rounds with Mike Tyson."

I replied, "I'm staying right here. I don't have it in me for another day like yesterday." So we lounged around, did some fishing and licked our wounds while we hoped for the wind to subside.

That night, a cold front came through. The rain came through my tent. But in the morning, the wind had changed direction and was greatly diminished. We loaded up and headed out, making for Spring Warrior Creek.

An Old Town canoe—a piece of excellence.

Two anglers use paddleboards somewhere in the Florida panhandle. *Courtesy of YOLOBoard/ Modus Photography.*

You're going to need protection from the sun when the day starts like this. *Courtesy of Mike Conneen.*

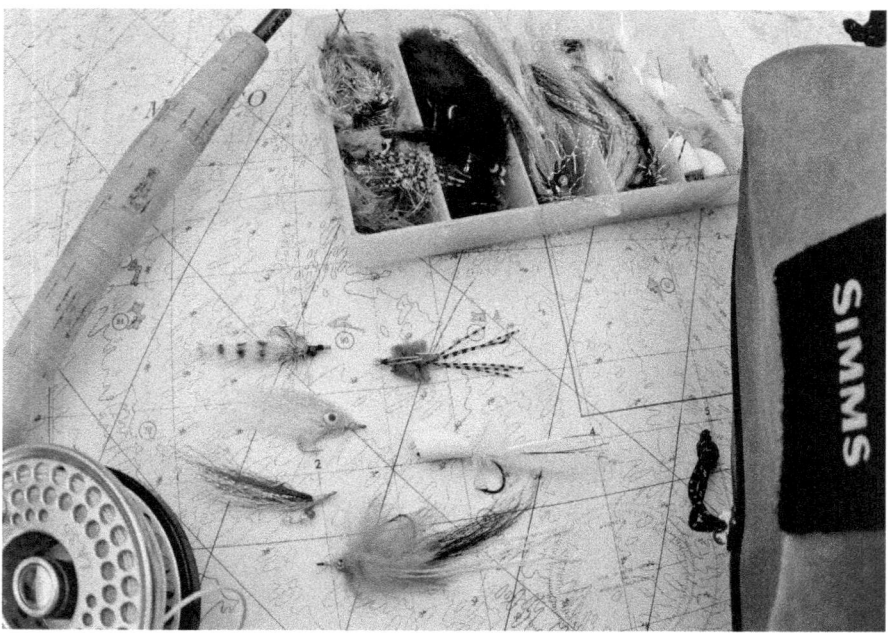

Be able to cover the water column with your flies. In saltwater, you need shrimp, crab and minnow imitations. In freshwater, nymphs and crayfish (not pictured) and minnow imitations will get you some fish. Poppers work well in both environments.

This spot will hold fish. There's a drop-off, a hole, cypress roots and fallen trees.

Bluefish are found statewide along the coast.

This crevalle jack is a common size inshore. They get to over thirty pounds, though.

Basin Bayou, a lovely place to fish.

A pair of oystercatchers—birds associated with oyster reefs.

The crystalline waters of St. Joe Bay make for fantastic sight fishing.

Sunset fishing along the Big Bend Paddling Trail near Rock Island.

A redfish tails in a flooded spartina grass marsh.

Mike Conneen got this nice redfish from Clapboard Creek.

Sunrise over the marsh at Canaveral National Seashore.

You'll find baby tarpon like this in both forks of Sebastian River.

An adult male bluegill from the Econ, fooled with a Road Runner.

The fish have spectacular colors in the clear spring water. This is a redbelly.

You'll find crappie in these creeks.

Right: Brian Stauffer got himself a flounder, too.

Below: Paddling past a waterfront property on the Chassahowitzka River. *Courtesy of Mike Conneen.*

Mike Conneen with a Hillsborough River bass.

Cooter turtles sunning themselves on Rainbow River.

Pine Island Sound is full of sea trout and other fish, too.

This cobia is a baby. There are some much larger ones in Pine Island Sound and Charlotte Harbor.

Everglades National Park has about one hundred linear miles of shoreline along the Gulf of Mexico and fantastic island campsites.

When fishing is good, sea trout like this will wear you out.

Launching or making landfall at low tide will be a problem in some places.

An anhinga in breeding plumage at the Anhinga Trail.

Above: A lower Deer Prairie Creek snook. *Courtesy of Mike Conneen.*

Left: Sailfish caught from a kayak in the Gulf Stream. *Courtesy of Denes Szakacs.*

Above: Mike Conneen shows off a red snapper taken in 150 feet of water. *Courtesy of Alex Gorichky.*

Right: Mike Conneen paddles around an alligator on the Myakka River.

We fooled this redfish with a DOA Shrimp.

As the tide bottomed out, the wind completely subsided. The surface of the Gulf slicked out. We enjoyed a leisurely paddle the rest of the way, casting jigs around creek mouths and points. The fish responded well. It proved to be the best stretch of fishing we would have on this trip.

This trip requires no fancy tackle. I brought a standard spinning outfit and a six-weight fly rod. Between the wind and the amount of paddling required, the fly rod was barely touched. If conditions were right, I suspect fly fishing could be awesome, though. Our weather killed us.

I caught fish on everything I tried (other than fly), but my most productive lure was a three-inch DOA CAL shad, either on an eighth-ounce weedless jighead or on a lightly weighted 3/0 worm hook. On lower tide phases, I could cast the lightly-weighted shad, and as the water deepened, I would switch to the jighead. While paddling, I'd just troll the shad behind me. This simple arrangement resulted in me catching quite a few fish. If you can cover the water column, which is not hard to do in four feet of water, and stay weedless (the grass is magnificently thick), you will do fine.

If, like we did, you use a sit-on-top kayak, you'll find keeping dry while paddling more important than tackle. I used chest waders and a Gore-Tex rain jacket, which worked well. Mike used rain pants and a rain jacket and just let his feet be wet and cold. Mike did not walk in the water like I did, though. Morning temperatures were in the forties on a couple of days, so it can and will get cold.

We took our trip in early December, at which time, prevailing winds usually come from the northwest. We had six windy days out of seven, and not one of them had northwest wind. That's simply bad luck.

Along this coast, at low tide, the grass flats are merely wet. You often paddle a half mile or more offshore in mere inches of water. There are rocks, oysters and spots that are merely damp. I had to get out and drag my vessel several hundred yards at one point.

If, like me, you use the ancient and honorable navigation system of map and compass, most of the time, you will have only the vaguest of idea of where you are. Everything looks the same, and much of the time, you are too far off the shoreline to see little details that might help you pinpoint location. If, like Mike, you use a smartphone with a GPS application and connect that device to a solar charger, you can find out exactly where you are any time.

While preparing for the trip, Mike had also gone to Google Maps, printed out a series of aerial photos of the route we were taking, collated and

Low tide was literally a drag. *Courtesy of Mike Conneen.*

marked them with the necessary information (where campsites and water were available, for example) and had them laminated. He was brilliantly prepared to navigate this stretch of coast. If not for him, I would probably still be there.

That solar charger was a slick device. For fifty dollars, it offers cheap insurance as far as making sure you end up where you intend.

We were paddling in inches of water about a half mile offshore. Of course, the wind blew. The water had just begun rising. I spotted a large hole, deeper water surrounded by practically dry land, and staked out next to it, as I was sure fish were there.

The first cast confirmed my suspicion. A sea trout, not quite large enough to hold batter, smacked the shad. A couple more casts produced another.

That hole provided the hottest piece of fishing the trip produced—twenty or thirty trout to twenty inches, a few redfish and a flounder—for about forty minutes. Then, the water got too deep, the bite dried up and we still had a long way to go. So, off we went.

Our last morning, we faced a nine-mile paddle to Mike's car, which was parked at Sea Hag Marina on the Steinhatchee River. We were going into the wind, but blessedly, it was light. The brilliant morning sun shone. Up ahead a half mile or so, I could see birds diving and what I thought were fish breaking. I paddled hard to get there, wondering what kind of fish they were, anticipating some hot action.

Silly me. What I thought were fish was a flock of mergansers chasing minnows, with their splashes in the glare of the sun and the diving gulls above them luring me into needless exertion. I didn't mind.

Steinhatchee came into view. It looked so close! I got off the kayak to take a short break and said to Mike, "The weather could not be any nicer!" He agreed.

I got back into the boat, took one stroke and BAM! Rudely, a harsh wind punched us right in the face, again. In an instant, it was blowing twenty. We had to go straight into it, with waves coming over the bow of the boat. Steinhatchee wasn't so close any more.

That last two miles took us close to two hours of struggle, the wind laughing as it toyed with us. We finally gained the lee shore of the river and worked our way back to Sea Hag Marina. Mike's truck sat there silently, waiting to take us home.

Nuts and Bolts

We took a sixty-mile paddle trip from the Aucilla River to Steinhatchee. We drove to Steinhatchee first, stayed in a cabin at Sea Hag Marina and got a shuttle the next morning to the Aucilla from Russ McAllister at Suwannee Guides (352.542.8331, www.suwanneeguides.com). There are other trips available and other ways to go about getting the shuttle done, but our method worked well and is recommended.

Camping not your idea of a good time? There's lodging available (bed and shower) along the way—not enough to eliminate camping, but enough to cut back on it.

Econfina River Resort on the Econfina River:
  850.584.2135, www.econfinaresort.biz
Spring Warrior Fish Camp on Spring Warrior Creek (these folks were very nice to us):
  850.838.2035, www.springwarriorfishcamp.com
Sea Hag Marina in Steinhatchee (these folks were really nice to us):
  352.498.3008, seahag.com

Rentals

River Haven Marina, Steinhatchee:
  352.498.0709, www.riverhavenmarine.net

The FWC issues camping permits for the sites along the paddling trail. They only issue these free permits to paddlers. You are allowed only one night at each site. Further, if you paddle a canoe, you can't get one—it's kayaks or no permit. I don't get it, but they make the rules.

At the following link, you can find information on obtaining permits, the trail guide, and further information. Examine the page carefully when you first go there. There's a lot of information, and some of it is easy to miss: myfwc.com/viewing/recreation/wmas/lead/big-bend/paddling-trail.

This trip is extremely rewarding. I hope you enjoy it.

History

During the Civil War (in Florida, you'll hear it called the War of Northern Aggression), Union gunboats tried to blockade Southern ports. The South has a long coastline that is hard to completely blockade. However, the blockade did significantly reduce the amount of trade. One of the commodities that quickly became scarce was salt.

The Confederate army used huge amounts of salt in salting pork for the troops. It was vital to the war supply, and homeowners needed it, too.

Many salt works were set up along the Gulf Coast. Men laboring in the salt works got deferments from military service, so there was no shortage of labor for this job. Although the biggest and best-known of these salt works was on St. Andrews Bay, there were a couple on the shoreline of Deadman's Bay, right by Steinhatchee

BULOW CREEK AND TOMOKA RIVER

These two streams lie within close proximity of each other just west of Ormond Beach. Both have similarities in that they start as small blackwater streams with bass, sunfish and other freshwater fishes and gradually become tidal and brackish with saltwater fish like redfish, trout, snook and baby tarpon. Sight fishing either stream is hard to do because of the water color and/or turbidity. Both are associated with state parks.

Bulow Creek has more and easier access for paddlers and an extensive salt marsh system that Tomoka River lacks. There are many nooks and crannies, and to my way of thinking, it's by far the better paddle fishing destination.

Tomoka River has two access points for paddlers: near the mouth of the river in Tomoka State Park ($, canoe/kayak rentals and camping available) and eight miles upstream at River Bend Nature Park. Adventurous fishermen can make a through trip. You'll need to set up a shuttle. Most of the way, you paddle through people's yards. Admittedly, they are nice yards, but they're still yards. There are quite a few powerboats, too.

On my most recent Tomoka River trip, I thrilled to the sight of a soaring hawk and saw the "footprint" of a manatee in the water. Fishing was slow (one micro-snook and one mudfish in three hours). I looked at people's homes and docks and thought about the incessant hum of internal combustion

Alex Kumiski with a Bulow Creek snook.

when a guy with a Weedwacker came into the yard nearest to me and started going at it only a few feet away. That was enough for me!

My son Alex and I recently visited Bulow Creek for the first time. At about 9:00 a.m., we launched two kayaks from the south side of High Bridge Road. Within minutes, a decent sea trout hit my Clouser Minnow. Shortly after, a shorty snook hit the same fly. In less than ten minutes, I'd already released two fish. It was warm, and fishing did not stay hot. However, we were able to get away from the road and most of the sounds of civilization (there is a small airport nearby), spending several hours in what appeared to be an off-the-beaten-path area. We did not see any other boaters.

While the fish we got were small, we ended up with five species between us. The water was dirty, and sight fishing is tough, but the birds were awesome, and the area was aesthetically pleasing. I've had worse days, especially at places I'd never been before. I'll be returning there.

Rentals

Tomoka Outpost, in Tomoka State Park (camping is available in the park): 386.673.0022, www.tomokaoutpost.com.

4
JACKSONVILLE AREA

You'll find miles and miles of spartina grass salt marsh laced with tidal creeks full of oyster bars and black pluff mud along the St. Johns River and Intracoastal Waterway. Paddlers have protected areas where they can fish for redfish, sea trout and flounder in most any weather. Fishing here is good! The Timucuan Ecological and Historical Preserve and Talbot Island State Park, north of the city, and the Guana Tolomato Matanzas National Estuarine Research Reserve on the south side offer beautiful, fish-filled areas to paddle. There are many other choices.

This area boasts the largest tidal range in the state, with the average range at Fernandina Beach being around six feet. When you fish here, you need to pay close attention to tide charts.

Friends who live and fly fish in this area generally prefer fishing an early, low, outgoing tide to sight fish for redfish. They fish through the turn of the tide until the oyster beds are covered with water, at which time their fishing day ends. At this point, they can't see the fish. Those fish are still there and can still be caught—the sight fishing, not the fishing, is done.

An exception to this happens every year in September and October, when the year's highest tides occur. Then, my friends wait for the marsh to flood and look for tailing redfish and sheepshead on top of the marsh in the thick marsh grass. It's something like fishing in a wet wheat field—quite unique and lots of fun when it's good.

The salt marshes in Jacksonville go on for miles. *Courtesy of Pixnio.*

CLAPBOARD CREEK

On a recent trip, Mike Conneen and I launched our boats at the Palms Fish Camp Restaurant boat ramp on Clapboard Creek. When we got there, the sun was not yet above the horizon. The tide was low and outgoing. We paddled across the creek and immediately found tailing redfish. I caught one on my first cast using a small, unweighted streamer fly. We had continuous shots at redfish all morning, with each of us getting several. Mike also got a couple of flounder. The farther away from Hecksher Drive you paddle, the less road noise you will hear.

Speaking of Hecksher Drive, there are many places along this road from which to launch kayaks, not the least of which are in Talbot Island State Park. You'll find Kayak Amelia on Simpson Creek, where you can rent a boat if you don't have one. Time it right, and you can catch an outgoing tide down to Nassau Sound and the incoming tide back to the starting point.

FORT GEORGE INLET

Recently, I launched at the Alimacani boat ramp near Fort George Inlet on a high incoming tide. There are sand shoals there that were covered up with redfish and black drum. These fish are right by a busy boat ramp and have seen lots of anglers. I tried to get a bite for a couple hours with both fly and spin tackle without success. However, you can catch the current up the Fort George River and fish other shoals and along the marsh grass.

GUANA RIVER

You'll find two paddle shops at Guana Preserve, the North Guana Outpost and the Guana Outpost South (owned by the same company). Both offer rentals, and they provide shuttles. The paddling distance between them is about eight miles through the Guana Preserve. This area is nontidal (there's a dam) but brackish. Fish species are the usual inshore types—redfish, sea trout, flounder, black drum.

Below the dam, you have the tidal fishing typical of the rest of the area, but it's very good. This whole area is beautiful, fishy and a great place to explore.

Nuts and Bolts

Visit www.jaxkayakfishing.com, an invaluable reference to anyone wanting to paddle fish in the greater Jacksonville area. The "Fishing Spots" link lists most of the places to fish around Jacksonville and includes launch information.

Rentals

Kayak Amelia, on Simpson Creek in Talbot Island State Park:
    904.251.0016, www.kayakamelia.com
North Guana Outpost:
    904.373.0306, northguanaoutpost.com

OCALA NATIONAL FOREST

Located in the middle of the state, the six-hundred-square-mile Ocala National Forest offers a variety of freshwater fishing opportunities to the paddler. You find the St. Johns River and Lake George on the eastern side and the Ocklawaha River (the largest tributary of the St. Johns) and Rodman Reservoir on the western side. In between are several spring runs (including Juniper Springs, Salt Springs and Alexander Springs) and about six hundred lakes and ponds. You can keep busy. The best paddling is probably on the Ocklawaha River.

OCKLAWAHA RIVER

This fast-flowing (for Florida) river, once a major steamboat route, suffered significant ecological damage during the twentieth century as a result of damming, dredging and rerouting. The dam resulted in the creation of Rodman Reservoir, a controversial (many people want the dam removed) but popular bass fishing lake. The Cross Florida Barge Canal, an ill-conceived project that would have dammed the river at two points and created two artificial reservoirs to facilitate navigation along the canal, was supposed to come through here. This project was defeated while under construction when Richard Nixon was president; however, the river is still blocked by Rodman Dam.

The Ocklawaha downstream from Gore's Landing has only light motorboat traffic and runs through undeveloped state land.

A warmouth from the Ocklawaha.

One of the nicest floats on the river runs from Gore's Landing to the State Road 316 bridge and can easily be done in a day. On a recent trip, we observed owls, hawks, otters, turtles and alligators. Even though the river was over its banks, we caught a few fish. Fish species you may encounter here include bass, crappie and various sunfish species.

Nuts and Bolts

Launch points on upper river: Ray Wayside Park (State Road 40 bridge), Gore's Landing, State Road 316 bridge (Lake Ocklawaha).

Launch points on lower river: Kirkpatrick Dam, State Road 19 bridge, or paddle onto the St. Johns River and pull out at Palatka.

You can overnight upstream of Rodman Reservoir.

Rentals

Ocklawaha Canoe Outpost:
352.236.4606, www.outpostresort.com

PELLICER CREEK

Like the other small blackwater creeks along Florida's east coast, Pellicer Creek rises in swamps west of Interstate 95. It drains east into the Matanzas River. Matanzas Inlet is two and a half miles to the north of the confluence of Pellicer Creek and the Matanzas River. Pellicer Creek is part of Florida's Designated Paddling Trail system. A map of the trail is available at floridadep.gov/file/9220/download?token=P1LNnC9U.

What this means to the fisherman is that one can theoretically catch black bass and sunfish and have a shot at tarpon in the same day without ever taking the boat out of the water. Redfish, sea trout, snook, flounder and bluefish are also distinct possibilities. The creek is even home to two species of sturgeon. Pellicer Creek was designed with paddle vessels in mind. The creek is shallow with extensive mudflats.

Pellicer is as wild a waterway as you'll find in this part of Florida. The state owns most of the riverbank between Interstate 95 and the Matanzas River, with Faver-Dykes State Park providing both camping and easy access for boaters. The straight line distance from U.S. Highway 1 to the Matanzas River is only about five miles, although the creek has many twists and turns in that distance.

On a hypothetical trip in a canoe or kayak launching at Highway 1 and heading east to the Matanzas River, you're advised to carry a couple of rods. One would be small for tangling with bass and bream in the upper section. Your usual sweetwater fly/lure selection works here, with sponge rubber spiders and small popping bugs providing plenty of action and exciting visuals.

Once you pass under Interstate 95 and the salinity increases, snook and redfish become probable. You may see them blasting bait along the margins of the creek, which widens. You'll need a larger rod. A selection of saltwater fly patterns including poppers, unweighted streamers and Clouser Minnows will be appropriate, as would surface plugs and soft plastic baits for the spin fisher.

The creek is shallow, but there are some deeper holes. Probe them with the sinking lures. In cold weather, these holes serve as thermal refuges and can load up with fish.

You'll pass the state park on the north bank. You can camp there or continue down toward the Intracoastal Waterway confluence, where you'll find the Pellicer Flats.

This large area of shallow water is loaded with oyster bars and surrounded by spartina grass. Slightly more than two miles from the ocean inlet, it's strongly affected by tides. Redfish, sea trout and flounder will be the primary targets, but surprises show up now and then.

If you're motivated, you can paddle up to the inlet. However, the Matanzas River is also the Intracoastal Waterway. Lots of large vessels use it, and this area is not recommended for users of hand-powered boats.

But that is no matter. Pellicer Creek is pretty, secluded and has the fish and solitude we seek. It's a lovely refuge from the stresses of the modern world.

Recently, I took a solo trip to Pellicer Creek, arriving about midday. Full of anticipation, I launched my kayak. It had been a while since my last visit, but fishing here had always been productive.

The place seemed dead. There were few signs of life. The water was still, with no current at all. Other fishermen I spoke with had not done much.

I paddled down to the Pellicer Flats, casting a plastic shad as I went, confident there would be fish around the plentiful oyster bars. But nothing was there. At least the water began to move as the tide started rising.

Utilizing wind and the current, I worked my way back toward the creek, casting as I went. Nothing. Three hours had passed since launching the kayak, yet I had not had a single strike.

A short distance into the creek was an obvious tide rip—not strong, but clearly defined. The first cast into it produced the first bite of the day, catching me so much by surprise that I missed it. On the next cast, I hooked a sea trout. While playing it, I dropped the anchor.

An hour went by while I caught fish after fish, both trout and redfish. There weren't any big ones, but it was pleasant getting bites.

Finally, it was time to go. Leaving the fish biting, I paddled back to the boat launch and loaded up. Persistence had paid off once again.

Two conclusions can be drawn from this admittedly small sample. The first, of course, is that persistence often does pay. Keep hunting for fish until you find some or run out of time. The second is that when fishing in Pellicer Creek, you want to be in moving water. At the slack tides, the fish just loaf.

Pellicer Creek sea trout were taking the soft plastic shad—it's good anywhere.

The moon was in a quarter phase during this particular trip. I have to wonder if the fishing improves around the time of the full and new moons. More investigation is needed!

The word "pristine" is loosely used around Florida to describe various landscapes, especially along waterways. People apply this word to Pellicer Creek. Before the Revolutionary War, Pellicer Creek was known as Woodcutters Creek. There was a sawmill here, and this entire area was clear-cut to provide oak planks and pine masts for British naval ships. The area is not exactly untouched by man.

The creek's current designation comes from a man by the name of Francisco Pellicer, a skilled carpenter and well-known citizen of St. Augustine. Around 1775, he established a farm on the creek that now bears his name.

Nuts and Bolts

You'll find two convenient launch points on the creek. One is at Faver-Dykes State Park ($) on the creek itself. The other is at Princess Place Preserve, located at the confluence of Pellicer Creek and Pellicer Flats. You can camp at either place.

You can do this as a through trip, but the creek is small and better suited for an out-and-back.

Rentals

Faver-Dykes State Park (canoes):
   904.794.0997, www.floridastateparks.org/parks-and-trails/faver-dykes-state-park

ST. MARYS RIVER

This stream forms the state line between Florida and Georgia and runs for 126 miles between the Okefenokee Swamp and Fernandina Beach. My experience (admittedly, limited) is that it's more of a paddling stream than a fishing destination. During my recent trip there, I saw no fish or reptiles and only a few birds. I did not get a bite. I cannot recommend it as a fishing spot, particularly in light of all the other excellent fishing in the greater Jacksonville area.

5
SUWANNEE AND SANTA FE RIVERS

SUWANNEE RIVER

When Mike Conneen and I told Steve at the Suwannee Canoe Outpost that we intended to paddle, camp and fish, he said, "You know that the Suwannee isn't the best place to fish in Florida." We did. He also showed us the U.S. Geological Survey (USGS) gauge reading over the past several days. The river had gone up nearly six feet because of heavy rains in Georgia. We wanted to go anyway, so he charged my card, and we made the shuttle.

There are fish in the Suwannee, including Gulf sturgeon. These beasts sometimes leap out of the water and have (only rarely) crushed boats and boaters. By law, you cannot fish for sturgeon. Less threatening species you can fish for include bass, sunfish and catfish.

The upper river has little vegetation. The fish-attracting structure consists of cypress tree roots, downed trees and limestone ledges. The lower river (below Branford) gets more vegetation but also gets a lot of motorboat traffic. While you can certainly fish (and camp) here, it is not recommended for paddlers.

The Suwannee comes out of the Okefenokee Swamp in Georgia, traversing nearly 250 miles before emptying into the Gulf of Mexico. Many springs along the way empty into the river. An important part of the paddle trail network in Florida, the Suwannee attracts lots of recreational paddlers. It even has a Class III rapid—Big Shoals. Camping is popular

Even when it's not fishing well, the Suwannee is a beautiful river.

along the river in the state parks (there are several along the river) and in the river camps.

The river goes up quickly after a rain, as Mike and I discovered. This will ruin the fishing. Mike and I had a lovely trip, but we did not catch a fish. I didn't try.

We paddled from the Suwannee Canoe Outpost to Gibson Park, a thirteen-mile stretch. We mostly floated on the current for a good part of the way. It was quiet (no road noise), the surrounding woods were regal and one could hardly imagine a nicer way to spend a couple of days.

The water is tannin-stained but clear. The banks are covered with tupelo, oak, pine, sweet gum and some cabbage palm. The Florida Trail runs for miles along the north/west bank. You can run the Suwannee from Fargo, Georgia, all the way to the Gulf of Mexico. The best stretch for paddlers runs from White Springs to Branford.

The state Suwannee River wilderness paddle guide is posted online at naturalnorthflorida.com/blog/wp-content/uploads/2015/09/State RoadWT_Guide2015_Ver2a.pdf.

Rentals

American Canoe Adventures, White Springs (this place is farthest upstream):
  386.397.1309, www.aca1.com
Suwannee Canoe Outpost, Live Oak:
  386.364.4991, www.suwanneeoutpost.com
Anderson's Outdoor Adventures:
  352.407.0059, www.andersonsoutdooradventures.com.
This place is farther down the river, south of Fanning Springs. It also offers rentals on the Santa Fe River.

SANTA FE RIVER

The seventy-five-mile-long Sante Fe River is the Suwannee's largest tributary. Attractive and heavily influenced by springs, it has vast areas of submerged vegetation in the middle and upper reaches. These areas harbor abundant freshwater shrimp, scuds and aquatic insects, which result in excellent growth rates for fish, particularly abundant redbreast sunfish and pugnacious spotted sunfish (stumpknockers).

The uppermost section of the river is not navigable because of blowdowns. The river becomes big enough to paddle where it comes out of the ground at River Rise Preserve State Park. Inside of O'Leno State Park, the Santa Fe River disappears underground. It hides down there for a couple of miles before popping up to the surface again in River Rise Preserve. From there, the state paddling trail runs twenty-six miles downstream to State Highway 129.

Between Highways 27 and 47, the river is popular with paddlers, tubers, motorboaters and divers due to several sizable springs. Avoid it during busy times.

Find the Santa Fe River Trail Guide at floridadep.gov/sites/default/files/Santa%20Fe%20Guide_0.pdf

Nuts and Bolts

Adventure Outpost:
  386-454-0611, http://adventureoutpost.net
Rum 138:
  386-454-4247, http://rum138.com
Santa Fe Canoe Outpost:
  386-454-2050, https://santaferiver.com

6
EAST CENTRAL FLORIDA

FELLSMERE GRADE RECREATION AREA

Fellsmere Grade Recreation Area (FGRA) gives access to two important pieces of bass fishing water: Stick Marsh, which has a boat ramp and is not recommended for paddlers, and Farm 13, which has no boat ramp (one is eventually supposed to be built) and is awesome for paddlers as long as it's not too windy. This is a big piece of open water with no place to hide from wind.

On the official map, Farm 13 is called the Fellsmere Water Management Area. It's easy to see why fishermen came up with a shorter name.

Kayak guide Dee Kaminski met Mike Conneen, Alex Kumiski and me out on the water on a spectacular late-fall day. She told us we were fishing too shallow and led us out to the "islands" (I suspect they float—they're certainly not something you can get out on), where we proceeded to catch some fish!

The lake is infested with hydrilla, an invasive water plant. In order to fish here, you need to have weedless lures. A plastic worm (or a fluke, lizard, weedless toad, etc.) with a single unweighted or lightly weighted hook is always in good taste. Kaminski likes to insert a rattle into her worm. Her doing that is a good reason for you to, as well.

Because so much of the lake is shallow, fly tackle can be extremely effective. I had good luck with both poppers and gurglers. I suspect a weedless streamer would be effective, too.

Farm 13 has a lot of aquatic vegetation.

Farm 13 produces some hefty bass, too.

DESTINATIONS

The fish we got were fat, healthy and beautifully colored. We got several in the five-pound range. Eight- and ten-pound fish are fairly frequently caught, usually around the canals that were dredged through the area before it was flooded.

FGRA is about two hours from Orlando, an hour from Melbourne and slightly more than an hour from Stuart. The fishing is worth the drive. Check it out.

Nuts and Bolts

No rentals here. You'll need to bring a boat. A kayak dolly will come in handy here. Because it's a lake, the trip is an out-and-back. The official map can be found here: www.sjrwmd.com/static/lands/recreation/Blue_Cypress_Conservation_Area.pdf.

Guide Dee Kaminski (the queen of Farm 13!):
 321.394.6874, www.reelkayakfishing.com

Dee Kaminski, the queen of Farm 13.

INDIAN RIVER LAGOON SYSTEM

The 160-mile-long Indian River Lagoon system averages three feet deep—it's a paddle fisher's paradise! Along the lagoon's bank, you'll find a national seashore, four national wildlife refuges, and five state parks. Three separate but connected lagoons make up the system: the Mosquito Lagoon, the Banana River Lagoon and the Indian River Lagoon. Five inlets, extensive flats, tributary creeks and seldom-visited backwaters give paddle fishers a wide range of opportunities for all kinds of saltwater fish.

For the past several years, water quality has been a problem in these lagoons. Algae blooms and freshwater dumping from Lake Okeechobee have had a negative effect on the habitat here. At the north end of the system, the Banana River Lagoon has been hit particularly hard.

You can still find great fishing, though. Due to the length of the lagoon, I'm going to break this into three sections. The upper lagoon runs from Ponce Inlet to Dragon Point, the middle section runs from Dragon Point to Fort Pierce Inlet and the lower section runs from Fort Pierce Inlet to Jupiter Inlet.

Tailing redfish in the Indian River Lagoon.

Destinations

UPPER SECTION

This section includes the entire Mosquito Lagoon, the north end of Indian River Lagoon and the entire Banana River Lagoon. You'll find lots of places to fish here, especially in the Merritt Island National Wildlife Refuge. You can launch anywhere you can get access to the water. Here are what I consider to be the best places.

CANAVERAL NATIONAL SEASHORE

At the north end of Mosquito Lagoon, a labyrinth of islands, channels and shallow ponds hosts loads of wading birds, dolphins, redfish and sea trout. Canaveral National Seashore ("the park") administers a dozen National Park Service campsites here. The park calls them wilderness sites, and they are only accessible by water. A few have a fire grate and a picnic table. The others have space for tents and that's about it.

I've gone camping here by canoe and kayak many times over the years. Sometimes, the fishing hasn't been good, but other times, it has been outstanding. Either way, I always have an enjoyable time.

Obtain the necessary camping permit ($) from Canaveral National Seashore's Turtle Mound station. Launch your boats at River Breeze Park in Oak Hill or at Turtle Mound on the east side of the lagoon.

After dinner, roast marshmallows and watch the sun set. After it gets dark, you can stargaze for hours. During the winter months, Orion (one of the easiest constellations to recognize), Gemini, Taurus and Canis Major are all clearly visible. Sirius, in Canis Major, is the brightest star in the sky! If you're lucky, you'll see satellites or a meteor streaking across the heavens. You certainly don't get to see meteors every day.

After breakfast, go fishing. Fishing being what it is, sometimes it's great, and sometimes you'll get skunked. But the wading birds will be thick. You will see pelicans, ospreys and maybe a bald eagle, and dolphins are common, too. If you don't catch any fish, cook some hot dogs over a blazing campfire. When you roast a wiener on a stick over an open fire, they are as good as hot dogs can be!

You can expect raccoons to visit your campsite while you're here. Make sure to pack your food in raccoon-proof containers. A hard plastic cooler with a rope tied around it works well. On one trip, we left our s'mores fixings, in a shopping bag, unattended on a table for less than five minutes. A raccoon found the bag and tried to steal it. The chocolate and crackers fell out as he ran off, but he made a successful getaway with all of our marshmallows.

The most common game fish here are redfish and sea trout. Our standard spinning outfit is appropriate. Effective lures include soft plastic jerkbaits, weedless gold spoons and small popping plugs. The area is shallow, with a lot of grass, so weedless lures are a must.

Popular natural baits include shrimp and cut mullet. Small crabs can also be effective.

I kayak fish this area a lot and use fly tackle. The preferred technique is to paddle along shorelines searching for fish to which to cast the fly. Particularly during the cooler months (when camping is most enjoyable), fishing for both reds and sea trout can be excellent.

Left: Raccoons may be cute, but they will tear up your gear and steal your food. *Courtesy of Pixnio.*

Below: A school of redfish cavorting in Mosquito Lagoon.

On the first camping trip my boys and I made here, we paddled over a school of at least two hundred redfish. We continued on to our campsite and dropped off our gear, then went right back to the fish. Between us, we got a dozen fish, many on fly tackle. If you spend some time hunting for fish here, you will usually find some.

Florida saltwater fishing laws apply here. All fishermen need a saltwater fishing license.

Nuts and Bolts

To camp at Canaveral National Seashore, you must obtain a permit ($) from the office at Turtle Mound. The park accepts reservations by telephone one week in advance, and during the busy spring season, reservations are recommended: 386-428-3384, ext. 10, www.nps.gov/cana/planyourvisit/reservations.htm.

Adjacent to Canaveral National Seashore is the Merritt Island National Wildlife Refuge. No camping is allowed here at any time. If you intend to camp on a spoil island in the Mosquito Lagoon, be sure to find the right one.

The Mosquito Lagoon wasn't named on a whim. The bugs can be nasty when the weather is warm, so the best camping time runs from Thanksgiving through Easter. The prudent camper will bring bug spray anytime they camp anywhere in Florida.

Rentals

JB's Fish Camp, New Smyrna Beach:
 386.427.5747, jbsfishcamp.com

History

On the Indian River Lagoon side of the Merritt Island National Wildlife Refuge, you'll find Dummitt Creek and Dummitt Cove. Douglass D. Dummitt, born in 1784 on the island of Barbados, was the son of an English planter. According to legend, Dummitt migrated to Spanish Florida in 1807 and was so attracted by the fragrance of orange blossoms along the east coast of Florida that he became determined to find the source. His search led him to the east bank of the Indian River Lagoon on the north end of Merritt Island, south of New Smyrna and near the present-day location of the John F. Kennedy Space Center. Here, he found a wild grove growing on high ground between two tidal basins, the Mosquito Lagoon and the Indian River Lagoon.

Dummitt established his own grove here—one of the earliest in the Indian River region. His first commercial fruit shipped in 1828. Dummitt's settlement also became one of the first permanent European settlements in the Cape Canaveral region of Florida.

The long Second Seminole War broke out in 1835. Unlike many others, Dummitt did not run from sporadic Indian raids. Dummitt was a loyal member of a Mosquito County militia called the Mosquito Roarers. He served as captain until he was wounded in the neck.

During his rehabilitation in St. Augustine, he married a socialite in 1837. That proved to be yet another constant pain in his neck. A handful of years later, she did something almost unheard of in those times—she filed for divorce. She simply wanted out.

After his rehabilitation, Dummitt resettled in Merritt Island to concentrate on oranges, with a brief stint as the Titusville area's first state representative after Florida gained statehood. He reestablished his grove compound between the two lagoons and mastered the grafting technique known as topworking: blending sweet orange buds onto the trunks of sour orange trees, rendering them resistant to frost that killed competitors' groves. His once-experimental oranges were also delicious, and northerners paid a premium for them. Dummitt supplied the demand for what became known as Indian River citrus.

Except for sharing his secrets of success with others in the area, Dummitt largely kept to himself in his waning years. He had three children with his longtime companion, a woman named Leandra Fernandez. In 1860, their teenage son tragically lost his life after being shot. This incident had a profound effect on Dummitt. The beaten-down old man sold his slaves at the start of the Civil War and died a recluse eight years after the war's end.

TURNBULL CREEK

Paddle trips most often are out-and-backs, in which you launch your boat, paddle out a way, then return to where you started. When Turnbull Creek isn't flowing much (which is most of the time), an out-and-back is easily done. But you can also make a through trip here, although it requires two vehicles. Begin by leaving a vehicle at Scottsmoor Landing. Then, launch your boats at the bridge over the creek at Highway 1 and paddle to the landing, which is about four miles away. You'll be paddling south, so this trip is best with a north wind.

Half the trip runs through the creek proper (it's small) and passes extensive (for this part of Florida) spartina grass marsh. Fish species include sea trout, redfish, snook (usually small ones) and tarpon (always small ones). You will see many birds, including spectacular roseate spoonbills, and you can also expect to see large reptiles.

After two miles of winding creek, you'll enter the Indian River Lagoon, heading south along shallow flats along the west shoreline. Scottsmoor Landing is an obvious boat ramp about two miles down the shoreline.

I love this trip. Relatively few people use this creek. Once you're away from the road noise of Highway 1, all you hear is the wind through the grass and the plop of happy mullet. It's beautiful.

Nuts and Bolts

Launch your boat at the Highway 1 bridge crossing. The northeast side is best. If you're doing a through paddle, leave your second car at Scottsmoor Landing. Do *not* leave anything valuable in this vehicle.

Rentals

Kayaks by Bo, Titusville:
 321.474.9365, kayaksbybo.com

BANANA RIVER LAGOON NO-MOTOR ZONE

Until about five years ago, this place had the finest light tackle inshore fishing in peninsular Florida. A series of algae blooms eliminated the seagrasses, and bacterial decomposition of that organic matter has caused a number of devastating fish kills. You can still fish here, and you may still catch a few fish.

But the sight fishing for snook, big redfish and big black drum is just about gone. Most days, you're lucky to be able to see the bottom. It's sad.

Access KARS Park ($), on the west side of the lagoon. Paddle north from there.

SEBASTIAN RIVER

Sebastian River offers an east central Florida fishery for tarpon, snook and other species and is wonderfully protected from all but the nastiest weather a mile or so west of the better-known Sebastian Inlet. The river has two branches, the canalized north branch and the meandering, natural south branch. Fish use both.

Launch your paddle vessel at one of two free public boat ramps on the south branch, Dale Wimbrow Park or Donald MacDonald Park, both on County Road 505 in Sebastian. Paddle east, looking for rolling tarpon, listening for popping snook and casting to likely spots as you go.

I seldom fish east of the railroad trestle. Always look for fish as you paddle. Tarpon in the ten- to thirty-pound class could be anywhere through here.

Just west of the railroad trestle is the best place in the river to hook up. The river is shallow right here, and the hooked tarpon jump like dervishes.

If nothing appears here, move up the north fork. Cruise quietly, parallel to the shoreline, about forty or fifty feet out, casting to or parallel to the shore. Look for rolling fish. In addition to tarpon, snook, redfish, crevalle jack and other species will be taken this way.

This fork is about twelve feet deep in the middle. Use a lure that sinks. The three-inch DOA Shrimp and the DOA Terror Eyz both work well. Fish them slowly.

Go until you see a sign near the dam that prohibits further entry. If you haven't seen or gotten strikes from tarpon by now, head back and search the much more scenic south fork.

Since this fork is relatively unaltered by man's activities, it has shallow bars and deep holes along the bends, and some time will need to be invested to learn them. Since the river has been posted as a slow-speed manatee zone, fishing pressure is quite light. It is much more attractive than the north fork.

A bluefish caught with a soft plastic shad in the north fork of the Sebastian River. *Photo by Cheryl Kumiski.*

You can also cruise the shoreline here or, alternatively, simply drift with wind or current. Cast blindly or wait for fish to roll and throw to them.

If you start having success, anchor or stake out and work that area until it cools off. Or drift through the hot spot, then paddle upwind or up-current and drift through it again.

After catching a tarpon, revive it until it can swim away. This fish could grow to weigh one hundred pounds and could give you a much bigger thrill someday. Please treat it like the treasure it is!

Nuts and Bolts

You can launch your vessel at two sites on the north fork: Dale Wimbrow Park, Roseland Avenue, or Donald MacDonald Park (camping available—$), also on Roseland Avenue.

Rentals

About Kayaks:
772.589.3469, www.aboutkayaks.net

TURKEY CREEK, GOAT CREEK, CRANE'S CREEK

You'll find other creeks between Melbourne and Sebastian River: the Eau Gallie River, Crane Creek, Turkey Creek and Goat Creek. All are accessible from Highway 1 or roads off of Highway 1. All offer baby tarpon, snook and other species. Mike Conneen, who lives in this area, had this to say: "Goat Creek is only lightly developed. There are few houses on it and zero businesses with the exception of a farm wedding venue. These creeks are mostly fresh water, and the farther you go upstream, the fresher the water gets. They all have high concentrations of Mayan cichlids (invasive), which some people like to target. You will also catch bass and gar in all of these right beside tarpon and snook.

"Goat Creek is short but also has the advantage of spoil islands just outside of the mouth where you can target the same tarpon and snook with the addition of reds and trout. Turkey Creek is the most scenic because it flows through the Turkey Creek Sanctuary. Most of these creeks have some depth and the best visibility is usually in the dry season, when there is less storm water flowing."

All these creeks offer protected areas to paddle when the wind howls and the Indian River Lagoon looks like a washing machine.

Speaking of the Indian River Lagoon, you can certainly fish in the lagoon itself. If the wind is not blowing, exploring around the spoil islands can turn up sea trout, jacks, ladyfish and other species. If the wind is blowing, fish the lee shoreline, paying particular attention to docks, oyster bars and any other structure.

The lagoon around the Sebastian Inlet offers tidal-dependent fishing for a wide variety of species. Given the strength of the currents that run through the inlet and the number, size and speed of the motorized vessels using it, paddling in the inlet itself is not recommended. But, areas to the north and south offer paddlers lots of fishing options. The canals known as Honest John's Canals, about six miles north of the inlet, are a well-known winter fishing spot for big sea trout. Honest John's Fish Camp is an excellent access point for paddlers and is the best place for up-to-the-minute information about fishing the area.

SOUTHERN SECTION

This section, from Fort Pierce Inlet to Jupiter Inlet, has been ravaged by polluted fresh water dumped from Lake Okeechobee. Thank you, Army Corps of Engineers! When the water is clean, fishing inside of Fort Pierce Inlet, inside of St. Lucie Inlet and in Hobe Sound can be excellent for snook, trout, tarpon, pompano, snapper, jacks and more. When they are dumping, the whole place is a dead zone. It's bad enough that several friends have moved away from the area. When the place you love to paddle fish turns into a dead zone for months at a time, what choice do you have?

I have not fished this area since 2013 and cannot recommend it due to the ongoing and severe water quality problems.

SPRUCE CREEK

No-see-ums buzz and bite, a minor annoyance, while mullet skip frantically, trying to evade toothy predators. A popper gets launched, hurtling through the air towards the panicked mullet. "Pop-pop-pop- BLAM!" What bugs?

Spruce Creek arises from a number of small streams that drain a swampy area between Interstates 4 and 95 west of Daytona Beach. Its mouth at the Intracoastal Waterway, about six miles from its beginnings in a straight line, lies less than a mile from Ponce de León Inlet and the Atlantic Ocean. In its twisted ten- or twelve-mile length, Spruce Creek dramatically changes character from a small, blackwater stream overhung by moss-draped live oaks and cabbage palms to a spartina grass–encircled and oyster bar–encrusted saltwater estuary.

I once fished here with Perry Young, in the tidal part of the creek, from a canoe. We saw bait being crushed along a shoreline and pulled the boat up on a bar so we could pursue the fish on foot. The tide was rising, and we didn't secure the boat. Next thing, I was swimming after our vessel after it floated away. We did get a couple jacks for our trouble, though, and Perry caught his first redfish ever using a popping plug.

Redfish on a fly at low tide at Spruce Creek.

Given the variety of habitats here, one might expect the creek to hold quite a variety of fish species. It does. Redbellies and other types of sunfish, largemouth bass and spotted gar roam the upper reaches. Snook, tarpon, redfish, sea trout, crevalle jack, flounder and more stalk the tidal zone. There is a crossover area where one cast might produce a largemouth and the next a snook. These are the types of dilemmas one begins to appreciate when one fishes in central Florida.

Due to the dark tannin color of the water, this is not a place for subtlety. Lures that flash, rattle, pop, gurgle, whistle, push water or otherwise attract attention work well. Quiet baits don't. Tackle needs depend somewhat on target species, but one needs to realize that the target species and the one that takes the lure may not be the same.

Sight fishing in the classic sense is difficult here. Ordinarily, one casts to structure—fallen logs, seawalls, docks, oyster bars. I always have my best success here when the predators I seek are visibly feeding, chasing bait, popping on shrimp, that sort of thing. Lower tide phases often offer better fishing.

Some anglers prefer fishing undercut banks with overhanging vegetation. Development along the creek has definitely changed the fish's hangouts through the years, though. Everyone who builds a house along the creek puts in a dock. There are some fine snook living under some of those docks. The best docks are located at bends in the creek. The water is deeper there, and the current flow carries bait into the fish, shaded and cool beneath the structure.

Other prime snook hangouts include the mouths of the smaller creeks along the main artery. Falling water forces the baitfish to abandon their hiding places in the side creeks, and the snook wait for the unfortunate minnows to be swept out to them.

Another location where you can try for snook is at the railroad trestle. Usually, it's dead during the day, but it can be great at night. Upstream of the trestle you find many houses. You can fish their docks for snook, but you'll be fishing in people's yards.

Fish the creek in the warmer months. With warmer days comes warmer water, and that makes snook and tarpon happy. Spruce Creek tarpon tend to be little fellows, to about twenty pounds, but Spruce Creek snook can reach respectable sizes, well over ten pounds. If you target these fish, you will need a bite tippet consisting of thirty-pound fluorocarbon.

Winter is the worst time to fish Spruce Creek. The water gets too cold. The mullet move out, and the predators follow. Sometimes, some big bluefish will winter in the deeper holes, but it's inconsistent fishing. Spring through fall can be good, though. Like anywhere else, the more you fish, the more you'll catch. It takes time to learn the fish's habits, and feeding locations change throughout the year. If you can find baitfish, you'll usually find some of the big boys around. They all come through here at various times of the year.

Around the time of the autumnal equinox, during the mullet run, the creek fills with bait. Spectacular fishing can be had then.

I have caught bluefish as large as eight pounds in Spruce Creek. I once had a Spanish mackerel free-jump into my boat here (I ate him for breakfast the next morning). We catch flounder here. The creek's proximity to Ponce Inlet creates a situation in which literally any saltwater fish species might show up.

The entire creek is a slow-speed manatee zone. As such, it's ideal for hand-powered boats like canoes and kayaks. Such boats can be launched from Highway 1 at Spruce Creek Park or from Cracker Creek Canoeing, a privately owned, well-shaded park at 1795 Taylor Road, Port Orange, west of Interstate 95. It's eight miles from Cracker Creek to Spruce Creek

Park, so a through paddle is possible. However, you will have to set up your own shuttle.

If launching on an out-and-back from Highway 1, plan your trip to catch the current one way, then catch the tide the other way to assist your return.

The river downstream of the railroad trestle is littered with oyster bars. They lack any kind of marking, making navigation hazardous. Use care, and you should be fine.

The Nature Conservancy found some of the shoreline of the creek important enough that it purchased 610 acres of rare bluff formations for use as a nature preserve. This tract, accessible by boat, contains a sand burial mound constructed by Native Americans between 1200 and 1500 AD. This area also boasts some magnificent scenery, and snook and largemouth bass can both be caught at the base of the bluffs.

Spruce Creek has further ecological significance, as it is the only home of the Spruce Creek King's Crown, a six-inch snail that preys upon the creek's oysters. Conservation groups work to protect the water quality of the creek to prevent the snail's demise.

You'll get some flounder in Spruce Creek, too.

Destinations

Central Florida's Spruce Creek is overlooked, almost forgotten and beautiful. Give it a try if you find yourself near here.

Spruce Creek is part of Florida's Designated Paddling Trails network. See the maps and paddle guide here: floridadep.gov/sites/default/files/Spruce_Ck_Guide_0.pdf.

Nuts and Bolts

You can launch your vessel off of Highway 1 in one of three places: At Spruce Creek Park, north of the creek, west of the highway; at Divito Park, south of the creek, west of the highway; and at a makeshift launch across from Divito Road, south of the creek, east of the highway.

You can also launch at Cracker Creek ($, 386.304.0778) in Port Orange, west of Interstate 95 on the fresher-water part of the creek. Cracker Creek also has rentals available. There is little motor-vessel traffic in this part of the creek. Cracker Creek is closed on Monday and Tuesday. The rest of the week, hours are from 8:00 a.m. to 5:00 p.m.

7
THE ST. JOHNS RIVER SYSTEM

Florida's longest river starts in (altered) marshes southwest of Melbourne and flows north to empty into the Atlantic in Jacksonville. Although it and its many tributaries offer superb opportunities for freshwater species, the river proper is heavily used by powerboats, especially fan/airboats. These vessels do not enhance the wilderness aesthetic that paddling anglers usually seek. Every highway crossing has a fish camp that offers airboat rides. They never stop and are especially odious on weekends. You've been warned.

THE SHAD RUN

American shad (and hickory shad, too) run up the St. Johns River every year to spawn after making an amazing journey to Florida from the Bay of Fundy in New Brunswick, Canada. The St. Johns is the southernmost river in which these fish spawn. The first fish are typically caught somewhere between Lemon Bluff and State Road 46 in late December. The run peaks in February and finishes around the middle of March.

This fishing centers on the stretch of river from Lemon Bluff in Osteen to near Tosahatchee Wildlife Management area in Christmas south of State Road 50. Early in the season, the Lemon Bluff area is best. As the fish push up the river, depending on the water level, the fishing moves with them. This changes from year to year, depending on the water level and the strength of the run. The number of fish entering the river varies widely from year to year.

Shad fishing on the St. Johns River can be great some years; other years, not so much.

Spin fishers use light or ultralight outfits with shad darts, crappie jigs, small spoons or some combination of these, often rigged in tandem. Trolling is often used to locate the fish. Once you know where they are, you anchor the boat and cast.

When there are a lot of fish around, you will often see them flipping at the surface. You'll see them chase minnows on top, too. They let you know when they're there.

Like spin fishers, fly-fishers use a variety of flies, from heavily weighted shad flies to nymphs to small streamers or even (rarely) small surface flies. While a floating line is usually preferred, sometimes a sink-tip or even a full sinking line works better. Again, it depends on the water level as well as the fish's behavior.

While fishing for shad, don't be surprised if other species take your lure. Largemouth bass, crappie, various sunfish species, sunshine bass and other surprises will all appear in your catch totals.

There is no online list of places to launch your vessel. You'll have to use Google to find these ramps: Lemon Bluff boat ramp, Osteen; Mullet Lake Park boat ramp, Geneva; C.S. Lee Park boat ramp, Geneva; and State Road 50 boat ramp, Christmas.

ECONLOCKHATCHEE RIVER

A blackwater stream and part of the state paddling trail system, the "Econ" begins its journey in Lake Conlin in the northern part of Osceola County. The river then flows north for fifty-four miles through Orange and Seminole Counties before spilling into the St. Johns River just south of State Road 46 in Geneva, passing through extensive cypress wetlands—the Econlockhatchee River Swamp—along the way.

If you could get access to Lake Conlin (surrounded by private property), you might be able to descend the entire river. It would be a feat comparable to, say, DeSoto's exploration of Florida. This may be a slight exaggeration, but I've never heard of anyone doing it. At this point, the river truly is a wild, if hemmed in, swamp covered up by brush and blowdowns, nearly dry at low water.

The first practical place to start an Econ trip is at State Road 50 in Orlando. On the north side of the highway and west side of the river, you'll find Hidden River RV Park. You can launch your paddle craft here for a five-dollar fee, which does not include a shuttle. For that, you're on your own.

You will still find blowdowns here, as well as numerous logjams. It's an obstacle course for paddlers. The lower the water, the more numerous the carries. But, the river is (barely) navigable by paddle craft and holds the fish the Econ is known for—largemouth bass, redbellies (redbreast sunfish) and stumpknockers (spotted sunfish). You'll also find bluegills, spotted gar, channel cats and bullheads and the occasional crappie. However, these species are more common in the lower reaches of the stream.

You'll hear jets, cars and chainsaws, but you'll also see hogs, deer and alligators. The river corridor is pressed on all sides by civilization, but it still remains remarkably wild. Moss-draped cypress trees grow around and in the river, giving it a tropical mystique.

We've reached a point in our narrative where we need to divide the stream into sections. I am somewhat arbitrarily dividing it into four pieces defined by road crossings. We'll call the first section, which is nearly inaccessible, the swamp. In it is everything north of State Road 50 in Orlando. Due to its inaccessibility, I have little to say about it.

The upper Econ runs between State Road 50 and State Road 419 in Oviedo. This section is small and overgrown, with lots of blowdowns. Deep holes provide hiding spots for fish, and all that lumber in the water provides plenty of cover. Other than by paddle craft, the stream is inaccessible, and

Going over a blowdown on the upper Econlockhatchee.

you can't access it with paddle craft when the water is low. Expect to exert yourself if you make this fifteen-mile trip. Expect it to take all day. You'll be going over, under, around and through many blowdowns and log jams. You won't see many other people.

The middle Econ runs from State Road 419 to Snow Hill Road in Chuluota, passing through the Little Big Econ State Forest. This is the most popular section of the river for paddlers. The surrounding state lands are open to the public, and hiking trails along the river offer fishing access to those without boats. This section runs for ten miles. Kind paddlers with saws generally keep the blowdowns cleared through here.

Unless it's been unusually dry, you should be able to float right through. Paddling without fishing this section takes four or five hours. A fishing rod will slow you down a lot, and fishing here can be good.

The lower section starts at Snow Hill Road and ends at the St. Johns River. If you're paddling this section, expect a long, eighteen-mile trip. Lots of folks launch small motor craft at C.S. Lee Park at State Road 46 in Geneva and run their boats up the river.

This is the place for catfish, if that's your game. All the other species mentioned above will be found through here, too. You'll see tilapia and plecostamus. There is shoreline access (if you're willing to hike or bike) through the Little Big Econ State Forest Wildlife Management Area.

This big Econlockhatchee redbelly struck a Road Runner.

There's a water-level gauge on the Econ at Snow Hill Road. This gauge is perhaps the most important tool a prospective fisherman can use when it comes to predicting potential fishing success. Experience tells me that if it reads over 2.0, fishing will likely be slow. We like the river running low and clear.

The waters of the Econ are dark with tannin, and the bottom is mostly sand. There is little rooted vegetation in the river. Most of the cover for fish here consists of lumber in its pre-cut form—trees and branches. You'll find lots of downed trees and branches! This affects the baits and lures you use.

I prefer fly fishing here. I tie all my sunfish flies on Aberdeen hooks. Once snagged on lumber, I can usually straighten this hook by gently and steadily pulling. When it bends enough, it pulls off the wood instead of the leader breaking and losing the fly. The hook can then be bent back into shape with pliers.

Those flies, as well as the bass flies, float. Not only are the strikes more exciting, in the upper and middle stretches of the river, you will get hung up

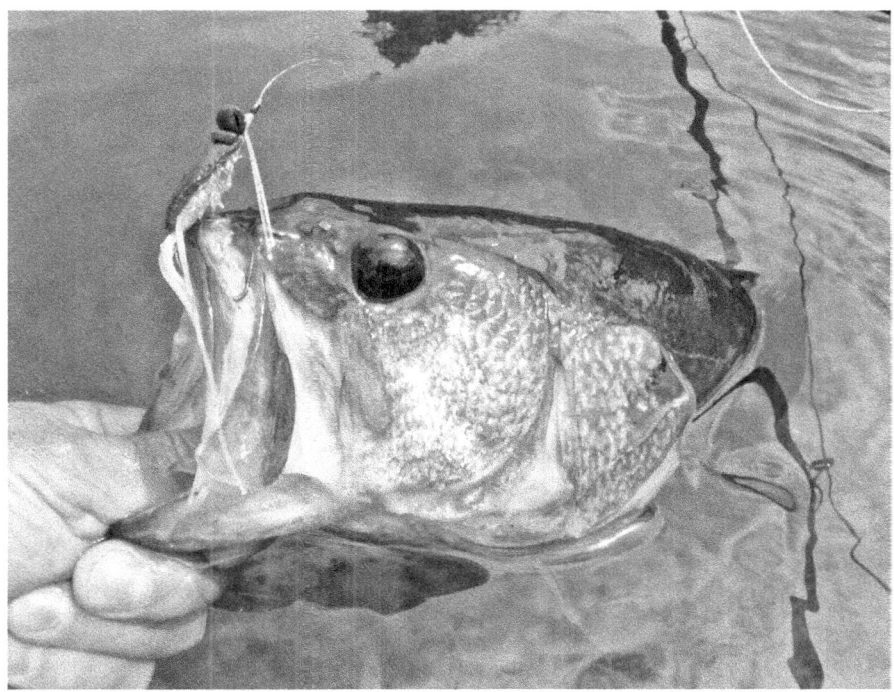

Fly fishing for bass on the Econ can be productive.

a lot less by using surface lures. When the water level is right and the fish are on, you'll have some good days.

One of my most memorable Econ fish was a big spotted gar. The bass bite on surface flies had been slow, and a streamer was tied on to the end of my leader. I had wade-fished a stretch of water and had just turned around to go back to my kayak when this brute of a gar crossed a shallow spot in front of me. Not expecting much, I dropped the fly in front of it. BOOM!

That fish struck that fly like a barracuda—a stunningly vicious strike. To my further surprise, the hook bit. After a spirited battle, I dragged the fish into shallow water, where I unhooked and released the critter.

Anglers using conventional tackle will find crappie jigs, micro-swimbaits, Road-Runners, small spoons and spinner baits and an assortment of soft plastic worms and lizards all work well. Again, the river lumber will rob you blind if you're not prepared. Think weedless!

My friend Tammy Wilson runs the Econ with friends three or four times a year, sometimes on overnight trips, usually on the lower part of

This gar struck like a barracuda!

the river. She had this to say about fishing there: "Bring a short rod. Fish every piece of structure you come across. Never go without at least one topwater lure or fly."

She went on to say, "Spinner baits are a must-have. Fished around downed trees, they have produced more and larger bass than any other lure during all of our fishing trips there. Gold spoon and natural or white skirts work the best, in the smallest size they make them. And never underestimate the power of a worm on a size six hook with a small split shot about a foot up the line."

While I'm out on the river, I run into other fishermen. There are two guys who fish for catfish in the lower part of the river using a Gheenoe. They assure me the best bait for big Econ cats is fresh chicken livers.

Other fishermen prefer freshly caught shad chunks (in season) or freshly peeled bait shrimp. Regardless of the bait, a one-half to one-ounce egg sinker rig keeps it on the bottom of the hole, usually at a bend in the lower river.

Speaking of shad, the confluence of the Econ with the St. Johns is a great place to fish for them. Many years, they swim up the Econ, sometimes as far as the bridge at Snow Hill Road.

Striped bass, or the striper hybrids called sunshine bass, also like the lower stretch of the Econ. I catch them incidentally while fishing for shad. You can tell the difference as soon as you hook up! Anglers target them, with peeled shrimp being a prime bait.

The Econ itself has one fishable tributary, the Little Econ. Unlike the mostly unaltered Econ proper, the Little Econ is extensively hydrologically altered, with substantial portions of the river channel canalized and interrupted by control structures. A number of canals draining various parts of the Orlando area flow into the Little Econ. You won't want to paddle here.

The pond behind the dam and the spillway below the dam at Jay Blanchard Park in Orlando are popular fishing areas on the Little Econ, though. Bass, catfish, sunfish and crappie are all found here. If fishing is slow, you can people-watch, especially on weekends.

You can find piscatorial excitement on the Econ, as well as gorgeous old-time Florida landscapes. In spite of the surrounding development, the river remains as beautiful and mysterious as its tannin-stained waters.

Econ Water Levels

As mentioned in the text, there's a river gauge on the Econ maintained by the U.S. Geological Survey. You can read the water level in real time here: waterdata.usgs.gov/fl/nwis/uv/?site_no=02233500&PARAmeter_cd=00065,00060.

After substantial rains, the Econ rises rapidly. Drainage ditches in Wedgefield in east Orange County feed into the Econ, and lots of drainage ditches in Orlando feed the Little Econ. The river can rise ten or twelve feet in a few days when the weather's wet.

Check the gauge before you go or suffer the slings and arrows of outrageous high water.

Nuts and Bolts

You are on your own as far as making a shuttle, and you will need to make one. You'll find Jay Blanchard Park (407.254.9030) on the Little Econ at 2451 N. Dean Road, Orlando, FL, 32817.

Launch and recover paddle vessels at these locations:
Hidden River RV Park:
  407.568.5346, five dollars per boat, access to upper Econ
Little Big Econ Canoe Launch: Access to middle Econ
Little Big Econ Forest Canoe Launch: Access to lower Econ and St. Johns River
C.S. Lee Park: Access to the lower Econ by way of the St. Johns River

ST. JOHNS RIVER SPRING CREEK TRIBUTARIES

Geology blessed central Florida with a number of spring creeks, including the Wekiva River and its tributaries (just minutes away from Orlando) and the Salt Springs Run, Alexander Springs Run and Juniper Springs Run, all in the Ocala National Forest. All of these creeks provide delightful float fishing by canoe or kayak on beautiful, crystal-clear waters for bream and largemouth in ecologically rich and natural environments.

Although each stream is unique, they share many physical characteristics. Each river begins as water wells up out of a hole in the ground, often a first magnitude spring. This water is clear, with a constant year-round temperature of 72 degrees Fahrenheit. Due to their mineral content, these spring runs support a lush growth of vegetation, including water lilies, bulrushes, dollar weed, cattails, pickerelweed, arrowhead, eelgrass and other plants.

All these plants provide a large, stable food base for a wide variety of animals. As you float, you'll be distracted from your fishing by an assortment of birds, including ospreys, turkeys, limpkins, herons, ibises and egrets. Otters, alligators, turtles and other animals are frequently nearby, too. Large numbers of minnows frolic in the shallows, and equally large numbers of aquatic insects, crayfish and other fish forage find refuge in the vegetation.

You might expect there to be large numbers of bass and bream to which you can cast. You would be right. Not only do these creeks hold respectable fish populations of their own, but fish also come into these streams from the St. Johns River to spawn, increasing fish populations on a seasonal basis. Striped bass also enter some of these creeks, wintering in some and summering in others.

You need a canoe or kayak to fish most of these streams. Wading is possible only on sandbars or where hard, white-sand bottom exists—for example, in some of the meadow sections of the Rock Springs Run. In most other places, thick deposits of sediment carpet the bottom, making wading difficult at best. Even these wadable sections are only accessible by paddle craft. There's no other way to reach them.

What tackle you need depends on what type of fish you prefer to catch. Fly-fishers will want a three- or four-weight outfit for panfish or a seven-weight for bigmouths. Floating lines are preferred.

A wide variety of flies can be used to take both bass and bream, and the only real difference between the flies used for these two species is the size of the hooks they are on. Fly-fishers will want flies on size 6 or 8 hooks for bream, and on size 1, 1/0 or even larger for bass. The bass frequently take

DESTINATIONS

Paddling on a beautiful day on spring-fed Rock Springs Run.

On spring creeks, the fish make up in numbers what they lack in size.

the smaller baits used for bream, and by using large baits, you miss out on a lot of the fun you'd have catching the smaller fish. I prefer using floating flies, but you should carry some subsurface patterns for days when fishing is slow.

Poppers, sliders, sponge rubber spiders and Muddler minnows make up a good, simple surface fly selection for bream. Rubber hackle on the spiders and poppers is important. If you tie your own flies, put weedguards on some of them to help reduce hang-ups. Bring extra flies, because you'll lose some to the vegetation. Sometimes, a big fish will burrow into lily pads and break you off, too.

For subsurface flies, some small Clouser minnows, wooly buggers, sinking rubber spiders and buggy looking nymph patterns like those used for trout will all be eagerly taken by hungry bream. Again, something to deter weeds will prove helpful, and extras will be needed to replace the inevitable losses.

A big selection of subsurface bass flies isn't needed. A couple of weighted crayfish imitations, a couple of streamers and some weighted eelworm patterns make up a small but effective selection. All of these flies need weedguards. They'll be worked in and around aquatic vegetation and fallen trees.

Spin fishers are best served with an ultralight outfit for bream and a standard spin outfit for bass. Spin fishers targeting bream will want lures like micro-plugs, Road Runners, Beetle Spins, crappie jigs and other similar small fare. For bass, jerk baits, weedless toads and plastic worms work well and can be worked through the thickest vegetation. A popper or other surface plug may come in handy.

Regardless of what kind of tackle you prefer, bring plenty of extras. You will lose some in the vegetation.

While you will sometimes get big fish, the bass in these streams tend to run small. There's no need to use heavy tackle and large baits unless you are truly only interested in big bass. Those large fish will take your little bream baits often enough to keep you interested in the fishing. The only problem with hooking a large fish on light tackle is pulling the big guy out of the weeds with that little rod after the hookup.

As you float down these rivers, the stream's character changes. You'll pass through open, meadow-like stretches with shallow water and thick growths of aquatic vegetation. These are followed by jungle-like stretches with the stream flowing through near-tunnels through the surrounding trees and lots of hanging vines. The water here is deep and dark. I just paddle through these jungle-like places. The streams here are narrow and overgrown, making casting difficult. I see few fish in these places, anyway. Why waste time fishing barren water?

The places to fish are the sunlit shallows. The fish thrive in the lush growth, where they easily find food. Work the edges and pockets in the vegetation as best as you can, paying particular attention to snags and blowdowns along the bank. Overhanging trees or cut banks often hold the biggest fish.

I prefer surface baits simply because the visuals they provide make fishing more fun. These are cast as close as possible to edges of any sort, current seams, drop-offs, docks, fallen trees or other structures and allowed to float freely with a minimum of "retrieve." Some experimentation is needed here, though. Some days, the fish want more action in their lures.

You'll find, as a rule, that casting accuracy is much more important than distance here. Fifty-foot casts are as much as you usually need, and most are shorter than this. In the clear water, the fish can see well. Keep your motion to a minimum.

A good sidearm cast that can skip a bait under overhanging vegetation is deadly. The more good water you cover, the better the odds of getting a strike.

You needn't cast blindly all the time. When the fish actively feed, you will hear them breaking the surface and see them jumping after dragonflies and other insects. I often stand in my canoe so I can see better. Sight fishing for bass is uncommon, but it can be done in these creeks. Wear earth tones or even a good camo pattern to avoid spooking the fish if you choose to use this technique.

Spotted sunfish (a.k.a. stumpknockers) are aggressive!

You can often see both bass and bream beds in the stream. Finding a collection of hundreds of active bream beds packed into a small area of stream is a subtle clue that fishing may be fast and furious for a while. Anchor a comfortable casting distance away, and work your bait along the outer edge of the beds at first, casting farther into the mass as fish on the outer beds get caught. This is bream fishing at its finest, with aggressive fish that hit on most casts.

A subtly colored paint job on the canoe is a good idea. Four of us recently floated the Alexander Springs Run in two canoes. Mine was a drab light green, the other one was bright red. We caught fish after fish; they caught nothing. There may have been other factors at work, but it's a good idea to be as discreet as possible with both your clothing and your watercraft.

Try fishing one of central Florida's spring creeks. You'll love the natural beauty of these streams and will find catching forty or fifty fish in a day a welcome novelty.

Nuts and Bolts

Wekiva River, seventeen miles long, has two canoe liveries, neither of which provides a shuttle service. You'll have to return the canoe to the place you rented it, which requires upstream paddling. The liveries are at Wekiva Springs State Park (407.884.4311) in Apopka and Wekiva Marina (407.862.1500) in Longwood.

Alexander Springs Run, in the Ocala National Forest off of C.R. 445, has a beautiful six-mile run. Paddle vessels are available for rent at the canoe concession (352.669.3522) in the National Forest Service Campground. They don't provide a shuttle service. You'll have to return the canoe to the place you rented it, which requires upstream paddling. Folks with their own boats can take out at 52 Landing six miles downstream.

Salt Springs Run, in the Ocala National Forest off of State Road 19. This is a five-mile run that terminates at Lake George. You'll have to return the canoe to the place you rented it, which requires upstream paddling. The phone number for the canoe livery is 352.685.2255.

Juniper Springs Run is a seven-mile run in the Ocala National Forest off of State Road 40. This creek is so small in the first two to three miles that fishing there is impossible. There is a canoe livery that provides a shuttle service. Call them at 352.625.3147.

DESTINATIONS

ROCK SPRINGS RUN

At this central Florida gem, it's more about the aesthetics than the size of the fish. You'll find plenty of brilliantly colored fish here, but they tend to be small ones—stumpknockers, redbellies and even the bass run small. Use ultralight tackle and small, weedless baits (I like a Beetle Spin or a three-weight fly rod), and you will have fun racking up the numbers, if not the poundage. A bruiser will surprise you on occasion.

The spring that starts the run rises in Kelly Park. Once the run leaves the park boundaries, it passes King's Landing (the put-in), a few pieces of private property and then winds eight miles through state park lands before joining the Wekiva River between Wekiwa Spring and Wekiva Marina (the take-out).

Those eight miles pass through wild country—the last remaining in this part of the state. You'll see deer, raccoons, turkeys, various types of reptiles and a wide assortment of birds, with an outside chance of spotting a black bear. Portions of the river pass through meadows resembling wildflower gardens. Primitive campsites are available (reservations required; www.floridastateparks.org/park-activities/Rock-Springs-Run Camping-Primitive).

Nuts and Bolts

Rentals, a launch and a shuttle are available at King's Landing (407.886.0859, kingslandingfl.com). Their employees will pick you up at Wekiva Marina and return you to the starting point for a fee.

8
WEST CENTRAL FLORIDA

ANCLOTE KEY

As you head north from Tampa Bay, Anclote is the last barrier island on the Gulf Coast of Florida until you get to Apalachicola. The state owns it and maintains it as a state park. No development! Primitive camping is allowed at the north end of the key right on the beach. It's a popular place, so make a reservation, and avoid the weekends!

You'll have to paddle out three miles (there's no bridge) and take everything you'll need with you (there are no stores, either). You'll also need to watch the weather forecast. Three miles of open water means you may get stuck out there waiting for the wind to subside. There are worse places to be stuck.

Fish species you may encounter here include the usual inshore suspects—snook, redfish, jacks and sea trout. You may also find Spanish mackerel, bluefish or cobia.

Nuts and Bolts

Anclote Key State Park:
   727.469.5943, www.floridastateparks.org/parks-and-trails/anclote-key-preserve-state-park
Best launch site: Fred Howard Park ($),
   www.pinellascounty.org/park/06_Howard.htm

DESTINATIONS

CHASSAHOWITZKA-HOMOSASSA-CRYSTAL RIVER

Because there's little shoreline development here, those in Citrus County refer to this coast as the Nature Coast. The paddle fisher will find mangrove forest, oysters, exposed limestone rock, myriad tidal creeks and extensive spartina grass marshes, with fish species being the usual inshore saltwater culprits: snook, redfish, sea trout, flounder, black drum, sheepshead, etc.

Paddle fishermen won't get in on this action, but the twelve-, sixteen- and twenty-pound tippet flyrod world-record tarpon were all caught in the Gulf of Mexico off of Homosassa. This is where Anthony Weston Dimock hooked that tarpon from a canoe over one hundred years ago.

You'll see lots of manatees. They draw a lot of tourists to the area during the winter, with Crystal River National Wildlife Refuge and Homosassa Springs State Park being the epicenters for spotting manatees. The Chassahowitzka gets its fair share of the blubbery mammals, too.

You'll also see plenty of fishermen. Some will be paddling, but many will be in motorboats, including airboats. Paddlers can minimize contact with motorboaters by fishing around rocks and oysters at lower-tide phases and doing something other than paddling on the weekends.

Thousands of miles of shoreline await the intrepid paddler. If ever there were a place where a GPS would be useful, that place is in the waters off Citrus County.

South of Homosassa and extending into Hernando County are the Chassahowitzka National Wildlife Refuge and the Chassahowitzka Wildlife Management Area, two sizable tracts of land with little remaining evidence of human habitation and mazes of islands and creeks.

The waterways between all the islands in the area tend to be shallow with deeper holes here and there. Any deep spot is likely to be loaded up with snook, trout and redfish during cold snaps. You'll find better fishing outside during the summer and inside during the winter.

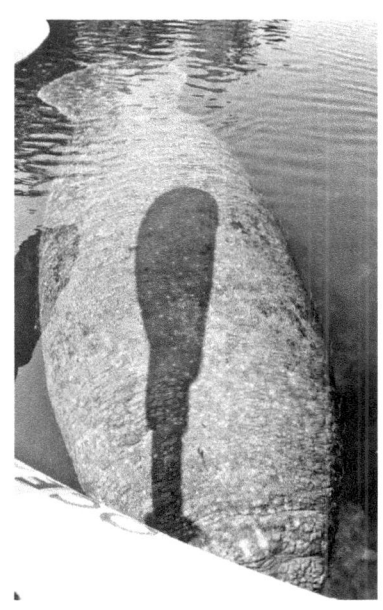

A curious manatee on the Chassahowitzka River.

Kayak guide Brian Stauffer with a redfish near Crystal River.

A recent fact-finding mission to this area included two days spent fishing with Brian Stauffer of Fish Head Kayak Charters. Both mornings, we launched out of Ozello Community Park. As so often happens, my timing was exquisite, with our fishing coming immediately after a cold front. (Note to reader: fishing immediately after a cold front passes is frequently a waste of time as far as catching fish is concerned.) We could see fish in the holes and sunning themselves on flats, but bites were hard to come by. Stauffer sent me a photo of the fisherman he took out the day after I left, who was holding a beautiful snook.

There were plenty of fish around—they were just in a post–cold front funk. Stauffer knows the area like his backyard, which it practically is, since he grew up near here.

Mike Conneen and I spent a day on the Chassahowitzka River, launching from the boat ramp at Chassahowitzka River Campground ($). Seven Sisters Springs is just upstream from the launch and is the jewel of the river, although there are other springs to explore. This is a popular spot. Get there early, and avoid weekends if you can. There are all kinds of boats using the river.

Fish species include snook, redfish, sea trout and flounder, along with jacks, ladyfish and snapper. They see a lot of fishermen and are not easy catches. You can paddle down into the salt marsh or, if you're ambitious and catch the tides right, you can go all the way to the Gulf of Mexico.

During our trip, we worked it hard, getting a pair of snook between us. Mike got some jacks and ladyfish where the marsh started. On the way back, with the sun up high, we could see all the snook we did not catch!

You will see some manatees here, especially around the springs.

Nuts and Bolts

Citrus County Visitors Bureau:
  800.587.6667, www.citrusbocc.com/visit/visit-bureau.htm
Brian Stauffer, Fish Head Kayak Charters:
  www.fishheadkayakcharters.com
Boat ramps:
  fishingkayaks.us/launch-sites, ocean.floridamarine.org/Boating_Guides/citrus/products/interactive_maps/map_side.html

Rentals

Riversport Kayaks, Homosassa:
  352.621.4972, riversportkayaks.com
Kayak Karavan, Chassahowitzka:
  888.821.0082, kayakkaravan.com
Chassahowitzka River Campground:
  352.382.2200, chassahowitzkaflorida.com

History

For thousands of years, before Europeans arrived, Native Americans hunted, fished and gathered wild plants along Florida's Gulf Coast. Evidence of several Native American campsites has been found within Chassahowitzka on the same high, dry ground used as camps by twentieth-century hunters. The swamp itself was doubtless as inhospitable a habitation site for Native Americans as it is for modern Floridians.

In 1528, the Panfilo de Narvarez expedition travelled north from Tampa Bay several miles inland from the coast, perhaps along the sand ridge bordering the eastern edge of the Chassahowitzka swamp.

Although no encampments or other sites have been found, the Seminole Indians were known to have been in the area during the Second Seminole War (1835–42). They gave the region the name Chassahowitzka, meaning "pumpkin-hanging place." This pumpkin was a small climbing variety that is now rare or perhaps extinct.

In the early 1900s, virgin bald cypress was harvested in the swamp. Southern red cedar, used to make pencils and cigar boxes, was logged after the marketable cypress was removed. A vast tram system was constructed for mules to haul timber from the swamp to a railroad in Homosassa. Many of the trams still remain and are now used by hunters, bikers, hikers, birders and nature photographers.

Between 1910 and 1922, Tidewater Cypress operated a lumber mill in Centralia, a town of 1,500 laborers and their families. The town had a well-stocked commissary, school, restaurant and even a theater, doctor and dentist. The mill, one of the largest in the state, could produce 100,000 board feet of lumber each day during peak periods. By 1938, the railroad ceased operations, and Centralia became a ghost town.

In 1985, land for the Chassahowitzka Refuge was purchased from the Lykes Brothers and the Turner Corporation as part of Florida's Conservation and Recreational Lands (CARL) program.

HILLSBOROUGH RIVER

Matt, Mike and I launched our vessels at Sargeant Park, off State Road 301, and paddled upstream a couple miles into Hillsborough River State Park. Leaves constantly fell, sometimes resembling snowflakes. The stream was small and clear with a slow, pleasant current. Fallen trees provided plentiful shelter for the plentiful bass. The river basin, heavily wooded, is subtly spectacular.

We took a break, then began floating back. Matt, tossing a surface fly, worked hard to get a bite on a chilly day. I was more interested in photography than fishing. But Mike, using a plastic worm, had a field day, catching bass after bass. Most were small, but a few were pushing three pounds. We saw some fatties hiding under fallen logs. This was good fishing for a late-fall day!

State Road 301 provided some road noise, but this is a gorgeous trip and highly recommended. Trips of several lengths are available. Canoe Escape is the best resource for further information.

Rentals

Hillsborough River State Park:
   813.987.6771. No shuttle available here.
Canoe Escape:
   813.986.2067, www.canoeescape.com. They provide a shuttle service. If you have your own boat, you can use Lyft or Uber to get a shuttle.

LITTLE MANATEE RIVER

The Little Manatee River is, well, little. The Little Manatee River Canoe Outpost has about 150 boats, and on busy weekends, they empty the lot. Avoid those busy times!

When we went on a weekday in early December, the place was deserted. We began by paddling upstream until the creek became impassable. Then, we floated with the current through Little Manatee River State Park to the take-out, spending most of the day doing so. It is five miles from the Canoe Outpost to the take-out. This trip was wonderful.

Plentiful wildflowers lined the banks and climbed the trees. Road noise was minimal. There were no houses until near the end.

The bass, including some lunkers, were biting on plastic worms. It was easy fishing in beautiful surroundings. This trip is highly recommended—if you avoid busy times.

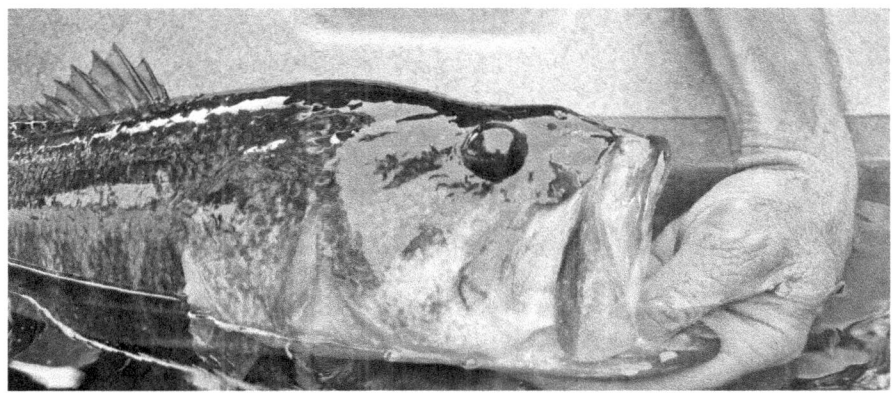

A fat bass ready for release back into the Little Manatee River.

Rentals

Little Manatee River State Park rents boats for out-and-backs only:
813.671.5005
Little Manatee River Canoe Outpost:
813.634.2228, www.thecanoeoutpost.com

LOXAHATCHEE RIVER

The Loxahatchee has an eight-and-a-half-mile state designated paddling trail, and is one of Florida's Wild and Scenic Rivers. It's a lovely place to fish! The upper part is fresh water, narrow and twisting, with cypress and cabbage palm trees lining the banks. The lower part is tidal, lined by mangroves. The state maintains an online paddling trail guide for this river at floridadep.gov/sites/default/files/Lox_guide_0.pdf.

One can fish the Loxahatchee as a through trip or do an out-and-back from either end. If you fish around the upper launch, be prepared for freshwater fish. If you fish around the state park, be prepared for saltwater

There are many jacks like this in the Loxahatchee River.

fish (and motor vessels). Through trippers need to be armed for bass and sunfish, snook and tarpon. Due to a couple dams and some blowdowns, the trip will take all day. Fortunately, you can get a shuttle (and rent a boat, if necessary) from Jupiter Outdoor Center.

My last Loxahatchee trip was with Mike Conneen. We launched kayaks at the state park and headed upstream. Although the weather was spectacular, it was an off day, fishing-wise. On the incoming tide, we found some small tarpon rolling. Nothing we tried (flies or various soft plastics) coerced them to bite. Nothing bit for several hours.

Once the tide started to fall, we found a cut near some homes that was packed with five-pound crevalle jacks. Mike got the first several using a small jig. When I switched to a similar bait, I started getting them, too. They were not what we had fantasized about, but this was much better than getting skunked. Crevalle have saved many a fisherman's day here in Florida!

Nuts and Bolts

Launch your vessel at one of two places: Loxahatchee Riverbend Park, upriver, the fresh water part; or Jonathan Dickinson State Park ($), downriver, the tidal part.

Rentals

Jupiter Outdoor Center:
561-745-6900, www.jupiteroutdoorcenter.com. Shuttle available.

MANATEE RIVER

My and Mike Conneen's day started by talking to a gentleman fishing at the boat ramp. A second, older gentleman, the father of the first, fished from the dock. Suddenly, the father was hooked up to a solid fish. He played it well! Moments later, I grabbed the jaw of a large snook and pulled it up onto the dock. It was an exciting beginning to our paddling day!

Unfortunately, we did not encounter any large snook. We did run into a large school of five-pound crevalle jacks, which provided photographic

You might catch a few fish if you find a school of jacks like this one!

and piscatorial entertainment for the better part of a couple hours. We got them on soft plastic minnows and fly rod poppers. Awesome! In addition to the snook and jacks, tarpon, trout, redfish and other inshore saltwater fish can be found.

Near Fort Hamer, the shorelines have homes, and there is motor-vessel traffic, but it's still a fine place to fish. As you move upstream, the river has a wilder aspect, and freshwater species become available. It's possible to do an overnight trip here, paddling from Ray's Canoe Hideaway to Fort Hamer Park.

The boat ramp we used is at the end of Fort Hamer Road, in Fort Hamer Park, near Parrish.

Rentals

Ray's Canoe Hideaway:
    941.747.3909, rayscanoehideaway.com

This picture offers a look at two different ways to solve the problem of how to catch fish.

History

Fort Hamer was a Seminole War fort—one of a chain of forts across Florida that were built around 1850. The army principally used Fort Hamer as a supply depot due to the accessibility of the port at the mouth of the river. Around this time, eighty or so Seminoles were convinced (and bribed) to leave Florida and move to Oklahoma. They were shipped out from Fort Hamer.

This fort was in service for twenty-seven years, after which time it was decommissioned and dismantled. Timbers from Fort Hamer were shipped south, where they were used in the construction of Fort Myers.

SARASOTA BAY

During the cooler months of the year, low tides tend to be quite low. All through southwest Florida, this is the best time to paddle fish. Low water cuts down on motorized competition.

Sarasota Bay offers a large body of water with lots of fishing opportunities. Passes at the north and south ends act like fish funnels, and the bars in the bay offer obvious targets for exploration. The northeast side of the bay has grass flats with potholes—always good places to look during low tides. The best areas usually have visible signs of baitfish, such as mullet.

Nuts and Bolts

Sarasota County Tourist Bureau:
  www.visitsarasota.com
Sarasota County launch sites:
  www.scgov.net/government/parks-recreation-natural-resources/things-to-do/facilities/boat-and-kayak-launches

Rentals

Sarasota Paddleboard Company:
  941.650.2241, www.sarasotapaddleboardcompany.com/longboatkey
Dolphin Paddlesports:
  941.922.9671, www.floridakayak.com

TAMPA BAY

In spite of being located in a major metropolitan area, Tampa Bay offers some great paddle fishing. It's big, and there are lots of places to launch a kayak. A few of the better ones are listed in this section.

Cockroach Bay Aquatic Preserve: Cockroach Bay Preserve State Park is a series of islands on the southeastern side of Tampa Bay. Two canoe/kayak paddling trails that are maintained by the Florida Coastal Office meander through the aquatic preserve that surrounds the islands. Habitats here include mangrove swamp, seagrass flats, oyster reefs and salt marsh. Fish species include snook, redfish, sea trout and more. The mangrove tunnels are cool! Motorboat traffic is fairly heavy along the Tampa Bay side of the preserve. Access is at the end of Cockroach Bay Road, 3839 Gulf City Road, Ruskin, FL, 33570.

Fort DeSoto Park (727.893.9185, www.pinellascounty.org/park/camping.htm): This park covers over one thousand acres with five interconnected islands near the mouth of Tampa Bay. This is some of the best fishing in the Tampa Bay area. Flats in the no-motor zone between the east end of the park and the Skyway Bridge are a great place to find spotted sea trout. Other spots, known as "bomb holes," can be excellent places to find redfish. Snook, tarpon, jacks, mackerel and more can all be found here. There is

The Snook Run trail at Cockroach Bay is well-marked.

also a 238-site family camping area with facilities. Make reservations! Rentals are available at Topwater Kayak Outpost (727.864.1991, www.unitedparkservices.com/kayak.html).

Weedon Island Preserve (www.weedonislandpreserve.org/paddling-launch.htm): On the St. Petersburg side of the bay is the Weedon Island Preserve. A five-mile paddle trail through mangrove tunnels and ponds in a no-motor area can be accessed at the end of Weedon Drive. For those who want to get off the main trail, there's all kinds of side trips through the preserve where you can lose yourself for a while. Fish species include snook, redfish, baby tarpon, sea trout and more. Consider going on a weekday, when the area is less crowded. Rentals are available at Sweetwater Kayaks (727.570.4844, sweetwaterkayaks.wordpress.com).

History

After the fateful 1492 voyage of Cristobal Colon, the Spanish Crown encouraged exploration and exploitation of the Americas and the Americans. La Florida was to be included as part of their empire. Several expeditions were sent to Florida in search of mineral wealth in the form of gold, silver and jewels; to obtain slaves; and to set up colonies to Christianize the natives.

Ponce de León led the first Florida expedition in 1513. This voyage only succeeded in mapping part of the Florida coastline; de León then returned to Cuba.

Ponce de León returned to Florida in 1521 intending to start a colony. Scholars argue about where this attempt was made. The natives violently opposed the Spanish. In a pitched battle with many fatalities on both sides, the Spaniards were repulsed, and de León was wounded. The survivors again retreated to Cuba, where de León died as a result of his wounds.

The year 1528 saw another Spanish attempt to colonize Florida, this one led by Panfilo de Narvaez. This expedition landed in the vicinity of Tampa Bay and succeeded only in teaching the natives about the Spanish (if they hadn't learned earlier). The men on this expedition were met with native attacks, illness and starvation. The only four survivors were rescued in 1536.

In May 1539, yet another expedition (consisting of five large and four smaller ships) led by Hernando de Soto set anchor just south of Tampa Bay. De Soto was a veteran conquistador who had gained wealth and fame from campaigns in Mexico and Peru. Within two days, they had located the bay, moved their ships into it and unloaded over seven hundred people and two hundred horses near the mouth of the Little Manatee River. Five hundred of these people were soldiers—bad news for the locals.

De Soto's army eventually marched north, looking for wealth. They made a fantastic trip, taking two years to travel through what is now Georgia, the Carolinas, Tennessee, Alabama and Mississippi before encountering the Mississippi River. They spent another year in Arkansas before they returned to the Mississippi River. Not only did they not find any gold, but De Soto died along the banks of the Mississippi in 1542.

The remnants of the army attempted to march to Mexico but soon realized the futility of this plan. They returned to the Mississippi River, where they constructed boats. The idea now was to float down the river to the Gulf, then sail to Mexico. In September 1543 the 311 survivors did in fact make landfall in Tampico, Mexico, ending this star-crossed expedition.

In spite of these colossal failures, the Spanish did manage to make a successful colony in St. Augustine, another in Pensacola, and built a string of missions along the Gulf Coast. In fact, the Spanish occupation of Florida lasted for over four hundred years. They successfully introduced citrus, feral hogs and horses to the peninsula, as well as measles, smallpox and other diseases. These illnesses caused raging epidemics among the natives, and by 1700, they had largely eliminated the people who lived here prior to the European invasion.

TENOROC FISH MANAGEMENT AREA AND OTHER PHOSPHATE PIT LAKES

Phosphate mining in Florida began in 1883. Early on, the mining was done by hand, and it was grueling work. As technology advanced, phosphate mining methods got more aggressive. Giant pits were dug into the ground by large machines to remove the fossils that contained the phosphate. Roads were built into the pit to allow access for trucks that hauled the mined material.

Pits fill with water.

The area that became Tenoroc was extensively surface-mined between 1950 and 1978 by the Coronet Phosphate Company, the Smith-Douglass Company, and Borden Inc. The name "Tenoroc" is "Coronet" spelled backwards. In September 1982, Borden Inc. donated 6,058 acres to the State of Florida. Two additional tracts were acquired through purchase: 341 acres (with funds from the Non-Mandatory Reclamation Trust Fund and Preservation 2000) in 1998, and 986 acres (through the Preservation 2000 Inholdings and Additions program) in 2000. In 2012, the Williams Acquisition Holding Company donated 770 acres to the State of Florida as an addition to Tenoroc.

Tenoroc currently exists as a mostly mined-over site, where lakes ranging in size from 7 to 227 acres provide quality public fishing. Two types—reclaimed and unreclaimed ("reclaimed" meaning all the islands and shoreline have been graded to create gradual slopes with deep water only in the center of cuts)—offer different fishing challenges. Both types offer open-water fishing opportunities, but their water depths and shoreline configurations greatly differ. Boat ramps are provided on most lakes. Some lakes also have restrooms and picnic pavilions.

Phosphate pits are fished differently than natural lakes. While bass and other fish still sometimes patrol the shorelines, they often hold over humps on the lake bottom, where old roadbeds or other structures were located. You can find these structures by trolling a deep-diving lipped plug or a crankbait. However, the easiest method is to use electronics in the form of a fish finder.

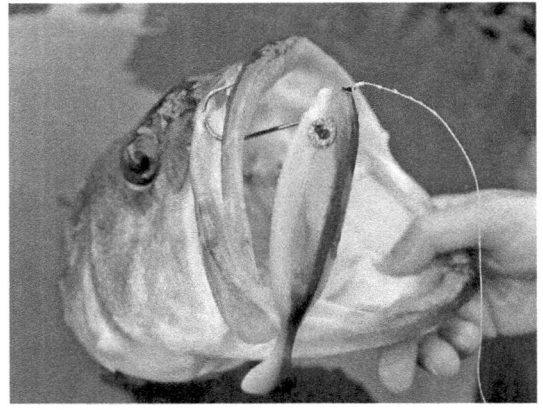

With the right combination of luck and skill, you can get bass like this from phosphate pit lakes.

For bigger bass, one of the best techniques is to cast to an underwater hump—water that's six or eight feet deep surrounded by much deeper water. The fish collect on top of those humps. Once you locate the hump, you cast an unpegged Texas-rigged worm (or substitute) out to where the hump is and leave the bail on the reel open until the bait stops sinking. That way, it falls straight down, not in an arc, as it sinks. As you crawl it back, you can feel the sinker bouncing off the rocks. On some casts, you never get to move the bait before a fish has it.

Tenoroc has twenty-three lakes. Some are reserved for special needs or shore-based fishermen. Many allow only hand-powered vessels and are ideal for paddling anglers.

Tenoroc (and Mosaic, a similar type of property) is open Friday through Monday. If you're visiting on Saturday or Sunday, you may want to call beforehand and make a reservation.

PITS IN POLK COUNTY

Tenoroc Fish Management Area, located two miles northeast of Lakeland. The odds of catching a quality bass and experiencing a peaceful fishing trip are high. No rentals are available here.
863.648.3200, myfwc.com/viewing/recreation/wmas/lead/tenoroc

Hardee County Park, Bowling Green, has four phosphate pits jointly managed by Hardee County and FWC:
863.767.1100, www.hardeecounty.net/site/content/parks/hardee_lakes.html

Destinations

Medard Park and Reservoir, a 770-acre reclaimed phosphate mine within Edward Medard Park located approximately six miles east of Brandon. Medard Reservoir is a fertile and productive impoundment with extensive, irregular shoreline.
813.757.3802, myfwc.com/fishing/freshwater/sites-forecast/sw/edward-medard-reservoir

Mary Holland Park, located just east of the Bartow Civic Center, has three interconnected lakes (Heron, Cardinal and Purple Martin). Fishing is allowed from the shore, piers and boats (no gasoline engines are permitted).
863.534.0120, blog.andythornal.com/blog/bid/38171/Fly-Fishing-Bartow-Mary-Holland-Park

Saddle Creek Park, a series of phosphate pits on 740 acres of mined phosphate land east of Lakeland:
863.499.2613, myfwc.com/fishing/freshwater/sites-forecast/sw/saddle-creek-park

Lake Crago Park, a 60-acre pit off State Road 33 in Lakeland, averages thirteen feet deep:
www.visitcentralflorida.org/m/destinations/lake-crago

Mosaic Fish Management Area, located about two miles south of Fort Meade, has high-quality reclaimed lakes and several mine-pit lakes:
863.648.3200, myfwc.com/fishing/freshwater/sites-forecast/black-bass/mosaic

PITS IN HAMILTON COUNTY

Eagle Lake, covering 200 acres, is an old and very fertile pit: myfwc.com/fishing/freshwater/sites-forecast/nc/phosphate-pits

Lang Lake Fish Management Area, an 86-acre lake, is a reclaimed pit myfwc.com/fishing/freshwater/sites-forecast/nc/phosphate-pits

WITHLACOOCHEE RIVER

Florida has two Withlacoochee Rivers! This section covers the southern Withlacoochee (the northern Withlacoochee is a tributary of the Suwannee and is also a part of the state paddle trail system), which runs 157 miles from the Green River swamp to the Gulf of Mexico near Yankeetown. It's part of the state paddle trail system. The paddle trail runs seventy-six miles from Lacoochee to Dunellon and holds the best scenery and fishing opportunities.

The paddle trail divides the stream into upper, middle and lower sections. I've only paddled the lower section from near State Road 200 to Dunellon. I'm particularly intrigued by this description of the middle section from the paddle trail guide: "The Middle Section is characterized by undefined banks and dense cypress swamp. Due to the confluence of the Little Withlacoochee River, the river becomes much wider and straighter. Homes are few and far between, and wildlife is abundant. The river is continually changing shape between wide, open lake sections and narrow, winding sections. There are few places to stop for rest breaks. There are a few fish camps that allow camping, and one public camping area at Shell Island with a couple of primitive campsites."

The Withlacoochee on an autumn day.

It sounds like a sweat-equity special—one I need to explore soon. The paddle trail guide details access and camping possibilities.

The stretch we paddled ran from Spruce Park, just downstream of State Road 200, to Dunellon—a nine-mile section we did in one day. Although the east bank was entirely wooded, there were many homes along the west bank. We fished it right after a cold front. We did not do much. I suspect that when the weather is stable, bass and sunfish bite a lot better, although the number of deck boats we saw makes me think the fish here are educated. Channel cats roam the river, too.

Like many other west-coast streams, the Withlacoochee was mined for phosphates. As you near Dunellon, several "lakes" open up along both sides of the river. The lakes are old phosphate pits. They're quite deep and are popular winter fishing spots for crappie.

Downstream a short way from Dunellon, the flow stops as you enter Lake Rousseau, an impoundment on the Withlacoochee. This impoundment has never been drawn down. Word has it that the bottom supports a thick layer of muck. A short distance downstream from the dam, you run into the unfinished Cross Florida Barge Canal. For most paddling anglers, the Withlacoochee ends at Dunellon.

RAINBOW RIVER

This stream, a tributary of the Withlacoochee, may have the clearest water in Florida. Millions of gallons of water come bubbling out of the ground every hour here—it's amazing!

With a state park and a county park on its banks and several liveries renting boats and tubes, it's a fisherman's nightmare on busy weekends and holidays. Even during slow times, you'll see other paddlers. While the east side is mostly woods, the west side has one house after another. It's not a place to go for solitude.

The Rainbow has plenty of fish, though. They can see everything you do and they are highly critical of most presentations. Mike Conneen caught a few bass on plastic worms by fishing them slowly, letting them drift with the current. I caught sunfish with a fly rod popper, casting upstream of where I thought the fish would be and letting the fly drift down to them. Use smaller lures and lighter leaders than you would anywhere else in Florida.

The headspring is in the state park. You can launch your boat here, but it's a hike from the parking lot. An easier place to launch is at **KP Hole County Park** ($). From there, it's a leisurely float back to Dunellon. Mike and I took our boats out at Rainbow River Canoe and Kayak, which offers rentals, shuttles and good advice.

Rentals

Rainbow River Canoe and Kayak, Dunellon:
 352.489.7854, www.rainbowrivercanoeandkayak.com
Withlacoochee River RV Park, Dade City:
 352.583.4778, www.canoegatorstyle.com
Withlacoochee Paddle Trail Guide:
 floridadep.gov/sites/default/files/WithSoGuide_0.pdf

9
SOUTH FLORIDA

Venice. Fort Myers. Naples. These towns—and the ones in between that were not mentioned—have several things in common: they are all along the Gulf Coast, they all have barrier islands near them and they all have heavy, heavy development. What this means to the paddle fishermen is that there are many boats in the water, and the fish are heavily fished. Don't expect the fishing here to be easy.

On the bright side, there are lots of restaurants and places to stay. The variety of fish is wide and includes trout, redfish, snook, tarpon, flounder, crevalle jacks, ladyfish, pompano and more. By picking the right times to fish, powerboats can be avoided.

The time to fish is at low-tide phases on full and new moons, when the water on the flats is too shallow for larger vessels and you can have the place to yourself. Snook and trout in potholes and redfish tailing in the shallows are a happy bonus at these times, too.

Wading is frequently the most effective way to fish the shallows here. Don't be afraid to get wet.

CHARLOTTE HARBOR AND PINE ISLAND SOUND

Use the Calusa Blueway as your paddle fishing resource through Pine Island Sound and Charlotte Harbor. Anthony Weston Dimock paddle fished this area for big tarpon—in the 1890s. In addition to tarpon, anglers will find snook, redfish, sea trout, flounder and many other saltwater species.

While there are numerous places to fish along the coast here, the undisputed gem of the area is the Charlotte Harbor/Pine Island Sound area. Charlotte Harbor stretches from the mouths of the Myakka and Peace Rivers some twenty miles all the way to Boca Grande Pass, intersecting there with the north end of Pine Island Sound. The sound is bordered on the west by Sanibel, Captiva, North Captiva and LaCosta Islands and on the east by the seventeen-mile-long Pine Island. They both have countless islands, mangrove trees, sand and oyster bars and miles of turtle grass flats filled with potholes. Snook, redfish and sea trout are the primary targets, but bluefish, ladyfish, Spanish mackerel, cobia, snapper, crevalle jacks, sharks and several other species will show up, often unexpectedly.

The principle baitfish in the sound are locally called whitebait. A three-inch soft plastic shad fished on a light jighead or a weedless hook imitates them well. Surface plugs also work well, especially on warm, calm mornings.

On the west side of the sound, Redfish, Captiva, and Boca Grand Passes separate the islands, allowing Gulf waters to pour into the sound on rising tides. Dimock fished for tarpon from a canoe all through here over one hundred years ago. Dimock may be gone, but the tarpon still return, with May and June being best months for the big ones. Dimock fly fished for them from a canoe, which is amazing. He and his partner ended up in the water a lot, which is not surprising at all.

During the colder months, the snook seek thermal refuges in creeks and canals away from the barrier islands. If you can find the right spot, some of the year's best fishing will happen during the year's most foul weather.

To the east of Pine Island, you'll find Matlacha (pronounced mat-la-SHAY) Pass. More fishing is available here all the way up the shoreline into Charlotte Harbor.

The Calusa Blueway Paddle Trail, 190 miles long, runs through Lee County, Pine Island Sound and Matlacha Pass. Find out more about it at www.fortmyers-sanibel.com/calusablueway.

I met my son Alex at Pineland Monument Park on an April morning at 9:30. We planned our trip around our mutually convenient times, not the tide. Naturally, the tide was high incoming—not exactly ideal for paddle fishing. In spite of this, we launched our boats and headed into Pine Island Sound, hoping to do battle with snook, redfish and sea trout.

Warm April sunlight fell on us. A gentle breeze ruffled the surface of the clear water. Bunches of turtle grass snaked toward the sunlight from the

Alex Kumiski with a redfish caught at Matlacha Pass.

bottom of the sound. Small minnows fled from our passage. Mullet jumped all around us.

Tossing a soft plastic shad into potholes in the grass, Alex hooked and caught a trout—not large, but the first of many we would get. As we worked the potholes, we'd get one here, one there—not hot fishing, but certainly steady. Some small ladyfish added variety.

Nearing some islands, I stood up to try some sight fishing. Although I was unable to get a cast to any of them, I saw four handsome reds and a couple small snook.

We found ourselves in Rocky Channel, the name of which I've never understood. There are no rocks, and it's not much of a channel. But there were deeper potholes here, and we cast our lures into them.

I got a bite, and line sizzled off the reel. Then, the fish came right in. It was a cobia that thought he was much bigger than he was. I took a photo, then we released him.

We continued casting as we worked our way back to the launch and continued to catch fish. Finally on dry land again, we loaded our boats and went looking for a restaurant.

The fishing hadn't been epic, but we'd gotten five species on a beautiful day and enjoyed some father-son bonding. What better way to spend a day?

DING DARLING NATIONAL WILDLIFE REFUGE

J.N. "Ding" Darling National Wildlife Refuge (239.472.1100, www.dingdarling.fws.gov) is located on Sanibel Island. It has a no-motors-allowed area—a wonderful place for paddlers. Snook, snapper, redfish, baby tarpon and sea trout are all available here. The best fishing days in Ding have two high tides and two or more feet of tide change. Days with minimum tide changes or those with only one high tide tend to be less productive.

As with the rest of Florida, in the winter months, passing cold fronts can quickly chill shallow waters and turn off the fish. On such days, it is easy to get skunked. A day or two after the front has passed usually brings calmer winds and warmer temperatures that get the action going again.

FORT MYERS HISTORICAL NOTE

Modern Fort Myers sits on the banks of the Caloosahatchee River. Before the arrival of the Spanish, this was the domain of the Calusa. Like most other Florida tribes, the Calusa were extinct by the time the Declaration of Independence was signed.

The Seminole, a tribe made up of renegade Indians from all over the Southeast as well as many runaway slaves, took up residence here. In addition to rich food sources, the two-mile-wide river provided ample opportunity to spot approaching enemies. The myriad feeder streams made disappearance into the Everglades easy.

After Florida became a U.S. territory, President Andrew Jackson wanted the Indians removed. A fort was built on the banks of the Caloosahatchee and named in honor of Colonel Abraham C. Myers; Fort Myers was used by the U.S. Army during the Seminole Wars. In 1865, the fort was decommissioned. Some of its timbers were used to build the first civilian structures in what is now downtown Fort Myers.

In 1881, industrialist Hamilton Disston came to the Caloosahatchee with the goal of "draining the swamps." He used dredging equipment to dig a canal from the Caloosahatchee to Lake Okeechobee. This allowed steamboats to travel from the Gulf through Lake Okeechobee and up the Kissimmee River.

In 1885, Fort Myers had a population nearing 350 people. Thomas Edison, then cruising the Gulf Coast, stopped in for a visit. He so liked what he found that he purchased thirteen acres along the Caloosahatchee and built a winter home and laboratory. Edison became a powerful force in the development of Fort Myers.

Henry Ford was one of Edison's friends. Ford also bought property in Fort Myers and promoted its development.

Today, the Edison and Ford Winter Estates (239.334.7419, www.edisonfordwinterestates.org) are operated as a single museum, which is open daily.

Nuts and Bolts

Lee County Tourist Bureau:
  www.leevcb.com
North Lee County launch sites:
  www.fortmyers-sanibel.com/media/16921/pineispaddlecraftlaunches.pdf
South Lee County launch sites:
  www.fortmyers-sanibel.com/media/16583/phase_1.pdf
Charlotte County Tourist Bureau:
  www.charlotteharbortravel.com
Charlotte County launch sites:
  www.saltchef.com/catch_fish/FL/Charlotte/boat_ramps.html
Collier County Tourist Bureau:
  www.colliercountyfl.gov/your-government/divisions-s-z/tourism/
    convention-visitors-bureau
Collier County launch sites:
  www.saltchef.com/catch_fish/FL/Collier/boat_ramps.html

Cayo Costa State Park, accessible only by water, boasts miles of undeveloped Gulf beaches, along with both tent sites and rental cabins:
  www.floridastateparks.org/park/Cayo-Costa
The Tropic Star can shuttle you out there and also has rental kayaks:
  239.283.0015, tropicstaradventures.com

Rentals

Charlotte County
West Wall Outfitters, Port Charlotte:
 941.875.9630, westwallflyshop.com
Downtown Bait and Tackle Shop, Punta Gorda:
 941.621.4190, www.downtownbaitshop.com/kayak-rentals

Lee County
Kayak Excursions, Fort Myers:
 239.297.7011, www.kayak-excursions.com
Tarpon Bay Explorers, Sanibel Island:
 239.472.8900, tarponbayexplorers.com
Estero River Outfitters, Estero:
 239.992.4050, www.esteroriveroutfitters.com
Carmen's Kayaks, Matlacha:
 239.333.7332, www.kayaktreks.com

Collier County
Naples Kayak Company:
 239.262.6149, napleskayakcompany.com
Adventures Kayaking, Bonita Beach:
 239.601.6976, www.adventureskayaking.com

EVERGLADES NATIONAL PARK

If you love wild places, the million-and-a-half-plus mostly roadless acres here offer the wildest place east of the Mississippi River. Dedicated by Harry Truman in 1947, the park features waters that offer great fresh- and saltwater fishing for those willing to work for it. Be prepared to camp!

Paddlers will find two entries into the untamed wilderness of Everglades National Park, a place every paddling angler should visit. These places are at Flamingo, an hour's drive from Miami, and Everglades City/Chokoloskee, an hour's drive from Naples. The truly adventurous paddler can plan a paddle/camping trip between these two points—a distance of about one hundred miles. Give yourself at least ten days (twelve is better!) if you decide to do this unless you want to paddle a lot and not fish at all. Regardless of

DESTINATIONS

The Florida City entrance to Everglades National Park.

how long or where you intend to visit, you should visit the national park website at www.nps.gov/ever/index.htm.

If you want a true survival test, visit during the summer. If you want to enjoy your trip, visiting in the months between Thanksgiving and Easter offers a more pleasant experience.

EVERGLADES CITY/CHOKOLOSKEE

There are motel rooms and restaurants in Everglades City. Civilization! Unlike in Flamingo, there is no campground here. If you want to camp, get into your vessel and paddle toward the Gulf into the Ten Thousand Islands. Several of these islands, including Tiger Key, Picnic Key, Jewel Key, Rabbit Key and Pavilion Key, have beaches on which you can camp and are (fairly) short paddles from either Everglades City or Chokoloskee. All have saltwater fish almost at your feet.

Just before Christmas, my son Maxx and I paddled out to Jewell Key from the Everglades City Ranger Station. We spent three days there fishing and

Snook like this are a real prize.

three nights stargazing. The sea trout fishing was stupid good, and we hooked enough snook to keep us on our toes. Shelling was fabulous, and we thoroughly enjoyed ourselves. And we did not need the first-aid kit, which is always a plus.

Like at Flamingo, there are a variety of loop trips you can take from here. Look at the park's Backcountry Planner (available on the park's website) to get some ideas.

History

On a sultry evening in October 1910, Edgar Watson, a homesteader living with his wife and three children on the Chatham River south of Chokoloskee, landed his skiff at the boat landing in that town. Every male resident of the town was there waiting for him. All of them carried guns. Some words were traded. Perhaps a threat or a threatening gesture was made. A few minutes later, Watson was lying in a puddle of his own blood with thirty-odd gunshot wounds leaking that precious fluid. Watson did not survive the incident.

The rumor was that Watson was on the lam. Why else would he have his family live through the heat and bugs of an Everglades summer? The rumor was that Watson hired helpers to work in his cane field—helpers who were never again seen on this earth.

The people of Chokoloskee were scared of Edgar Watson. Exactly what happened that evening and exactly why it happened have been the subject of lots of speculation. The best of that speculation was done by Peter Matthiessen in his book *Killing Mister Watson*. If you want to know more about the history of this part of what's now Everglades National Park and why Edgar Watson was a man to be feared, read this book.

Nuts and Bolts

Lodging: In Everglades City, you can find lodging at several places. No camping is available here other than in the backcountry (permit required).

Restaurants: You will find several restaurants in Everglades City. The Oar House has been there a long time.

Fishing/Boating/Camping supplies: You'll find several small stores in Everglades City and quite a selection of fishing tackle at the Outdoor Resorts store on Chokoloskee Island.

Paddler's access: You can launch your vessel right behind the Everglades City Ranger Station or at the boat ramp at Outdoor Resorts on Chokoloskee Island.

Guides: Captain Charles Wright, of Everglades Kayak Fishing (www.evergladeskayakfishing.com), offers kayak mothership trips into national park waters. He transports you and your paddle craft in a large Carolina Skiff, then drops you off. He waits while you fish, then transports you back to Everglades City.

Rentals

Ivey House:
  239.695.4666, iveyhouse.com
Outdoor Resorts:
  239.695.2881, www.outdoorresortsofchokoloskee.com

In the fall of 2014, my good friend Mike Conneen and I were camped on a beach near Cape Sable, west of Flamingo. At low tide that afternoon, the water's edge was way off the beach, leaving a couple hundred yards of sticky, wet mud between us and the water. We decided that by getting up at about 3:00 a.m., we could catch high water and make our getaway.

Mike has difficulty walking, so I gathered firewood. The wood was old, with loose, crumbly bark. As I carried it to our campsite, I thought, "This wood has scorpions in it."

As we sat by the campfire that evening, Mike broke the pieces of wood across his leg before feeding them to our fire. I warned him about the possibility of scorpions. Florida scorpions are small, a couple of inches long. You wouldn't see them in the dark.

After a pleasant evening watching the flames and the rising moon while conversing, we retired to our respective tents.

What seemed like just a few minutes later, I awoke to the sound of tent poles banging together. I checked my clock—it was 11:30. I looked out of my tent. Mike, quite visible in the moonlight, was taking down his tent.

Curious, I got up and walked over. "What are you doing? I thought we said three o'clock."

"A scorpion stung me." He showed it to me. "I can't go back in the tent. There might be another one."

I considered this. There's no argument against impeccable logic. I wouldn't want to be stung by a scorpion once, never mind twice, especially in the middle of the night in the supposed security of my tent. I turned around, went back to my tent and started packing up.

By 0030, we were packed, with our kayaks loaded up and floating at the water's edge. We climbed in and shoved off. I said to Mike, "I've never been stung by a scorpion. Where did it get you, and what does it feel like?"

He replied, "It got me in the back of my shoulder. It hurts like a hornet's sting, but it's not going away."

The surface of Florida Bay was unruffled by any air movement—a perfect mirror for the not-quite-full moon and the stars. Except for the mullet jumping here and there, an occasional mosquito buzz and the quiet splash of our paddles, it was quite silent.

The moon set. The Milky Way appeared in all its glory. Satellites and meteors streaked across the heavens. It was spectacular. I don't often paddle by starlight, but I recommend it from an aesthetic viewpoint.

We got to the dredge hole behind Flamingo Campground at first light. Boats still laden with camping gear, we started fishing, connecting with crevalle jacks, ladyfish and a few sea trout. Some tarpon rolled, but they ignored us.

As the giant ball of flame appeared on the eastern horizon and climbed into the sky, the fish stopped biting. Tired and sleepy, we paddled the rest of the way to Flamingo.

Destinations

FLAMINGO

The paddler won't find any lodging in Flamingo. You can camp in an RV or in a tent, but if you want to stay here, you'll be camping. The mosquitos here can be as bad as you will ever see. Be prepared by wearing long sleeves and pants and having a 25 percent DEET repellant spray. Spray the DEET on your hat and clothes and on a bandana you can tie around your neck. If that's not enough, wear a head net.

If you've been around mosquitoes much, you know they like some folks more than others. They find me tasty, so I have to resort to all the techniques when they're bad.

Flamingo offers freshwater fishing for bass, sunfish and exotics in the many ponds along the main park road. These ponds are lightly fished, and as a result, the fish are dumb. Fishing can be great!

In Flamingo proper, there is a boat ramp on the Florida Bay side. This ramp gives access to the flats and channels of Florida Bay and Snake Bight. All the popular inshore saltwater fish can be caught here, and the tarpon can be huge. My personal best snook was caught on a plug (a Bagley's Finger Mullet) in the dredge hole behind the Flamingo campground. This hole gets jacks, ladyfish, sea trout and tarpon in it, too.

The ramp on the Buttonwood Canal gives access to the Whitewater Bay side by a long paddle through said canal. Launch at Coot Bay Pond to shorten the trip into Coot Bay and Whitewater Bay. When you launch here, look for fish immediately. Baby tarpon—true juniors, a foot or so in length—love it in here. Tiny surface flies like gurglers are the best lures. You may find a stray redfish or snook in here, too.

Other saltwater fishing spots include West Lake and the Hell's Bay Canoe Trail, where bass and exotics are also a possibility.

The paddle trailhead at Hell's Bay

All kinds and lengths of overnight loop trips are available through here, using Flamingo as your base. Some examples:

- Launch at West Lake, paddle across the lake, through some creeks and ponds, out into Garfield Bight. Camp on the Shark Point chickee (seabirds roost here; be prepared for guano). Paddle across Snake Bight back to Flamingo. You'll have to beg a ride back to West Lake to recover your vehicle. You can do this as an overnighter, but two nights give you some fishing and exploration time.
- Launch at Coot Bay Pond, paddle up into Whitewater Bay, though Shark River and down the Gulf coast back to Flamingo. You need six or seven days to do this properly.
- Launch at the Hell's Bay Canoe Trail; go through Pearl Bay, Hell's Bay and Lane Bay and down the Lane River to Whitewater Bay; down the Shark River to the Gulf; then south down the coast to Flamingo. This requires the highest level of navigation skills and will take nine or ten days. You'll have to beg a ride back to the trailhead to recover your vehicle for this trip, too.

In my opinion, camping at Middle Cape Sable is an experience not to be missed, as it is the most fantastic spot in the southeastern United States. You can do this as an out-and-back. It's about fifteen miles each way. The weekend traffic is considerable. Try to visit during the week.

Many saltwater fishing day trips are available here, too. The no-motors-allowed area is a wild play-place for paddlers. If you find yourself in this part of the park, do the tourist thing and visit Mahogany Hammock, the Gumbo Limbo trail and the Anhinga Trail. You will be glad you did.

When my sons Maxx and Alex were thirteen and eleven years old, respectively, we took a nine-day, three-canoe trip in December with Ken Shannon, Mark Lolly and Mark's twelve-year-old son Jason. We started at the Hell's Bay Canoe Trail and made the trip described above. As we camped on the Hell's Bay chickee the first night, we were sitting in the dark, quietly conversing, when Alex got up to take care of some business. His excited voice soon cut through the night air: "Guys! I'm peeing on an alligator!" I'm glad he didn't fall off the chickee.

A few days later, we had a long paddle to get to the Joe River chickee. When we got there, it was warm. The tide was high. Ken likes to swim.

Within minutes, Ken and the three boys were jumping in the chickee, splashing around and having tons of fun.

Then, Jason sliced his leg on an oyster growing on one of the pilings. The cut was deep, running from just below his knee to just above his ankle. We had been paddling for five days and were as far away from help as we would get on this trip, and now we weren't having fun.

I got the first-aid kit out, and Ken cleaned the wound and patched it. He told Jason, "Don't let it get wet!" While that's hard to accomplish on a canoe trip, Jason did keep it dry for the duration.

A couple days later was Christmas. We paddled all day, getting to Cape Sable just after sunset. Within minutes, Alex had caught a twenty-pound blacktip shark. Within minutes, it was in pieces, frying in a pan. It wasn't the best fish I've ever eaten, but it certainly hit the spot after a long day!

A couple days later, we were back in Flamingo. Jason's wound was healing nicely. (Today, he has a long, white scar where the cut was.) It had been a great, rewarding trip—boys paddling, fishing, making fires, watching stars and satellites and making lifelong memories. Taking children paddling, fishing and camping is highly recommended!

History

In Florida Bay, close to Flamingo Campground, is a small island called Bradley Key. Beginning in the late nineteenth century, women's fashion dictated the wearing of bird's feathers, usually on hats. The most popular feathers were those from breeding egrets, but all kinds of wading bird's feathers were used.

Plume-hunters would go into bird rookeries and shoot the birds off their nests, pull off the feathers and leave the carcasses to rot and the eggs and fledgling birds to die. Millions of birds were slaughtered each year, and several species faced extinction.

In 1902, Florida passed a law to protect the birds that remained. They needed a warden to enforce this law, and Guy Bradley got the job.

Bradley took his job seriously, to the annoyance of many of his neighbors. He was constantly threatened and often heard bullets whizzing by his head.

In 1905, Bradley tried to arrest some plume-hunters near his home in Flamingo. One of the hunters shot him. His body was found the next day in his skiff, which floated near the island we now call Bradley Key.

Bradley's murder, and the murder of two other game wardens soon after, prompted the state of New York to outlaw the trade in wild-bird plumes in 1910. The fashion craze for feathers eventually dwindled. Egrets and other wading birds are now protected by law.

Guy Bradley was buried on a low ridge on Cape Sable.

Nuts and Bolts

Lodging: There's a campground in Flamingo and backcountry campsites (permit required). The nearest lodging is outside the park in Florida City.

Restaurants: There's a hot dog stand underneath the Flamingo Ranger Station. Otherwise, the closest restaurants are outside the park in Florida City.

Fishing/boating/camping supplies: There are grocery stores in Florida City. The Flamingo Marina store has basic supplies for fishermen and campers; there's also a gas pump there.

Paddler's access: Boat ramps are in Flamingo on the Florida Bay side and on the Buttonwood Canal. You can launch paddle vessels at Coot Bay Pond, West Lake, the Hell's Bay Canoe Trail and several other spots along the main park road.

Shuttle service: The Ivey House (iveyhouse.com) in Everglades City offers a shuttle service for those who want to paddle the length of the park.

Rentals

Flamingo Marina has canoes and kayaks for rent, with restrictions on where you can take them:
305-501-2852, evergladesadventures.com

MYAKKA RIVER

The Myakka River originates as a tiny stream north of State Road 64 in Manatee County. It is part of the state paddle trail system. It winds for sixty-six twisty miles, entering Charlotte Harbor at El Jobean. Along the way, it flows through two lakes in Myakka River State Park, one of Florida's oldest state parks, with fifty-eight square miles of wetlands, prairies, hammocks and pinelands.

Above Myakka City, the river is just a trickle through a swamp of deadfalls, dense thickets and barbed-wire cattle fences. It's not much different downstream. A dedicated, motivated paddler could fight his or her way down to the County Road 780 bridge, but it won't be easy. At said bridge, launching is reported to be possible but not easy—this should only be attempted by a dedicated, motivated paddler.

The first practical place to launch a boat is on the upper lake in the state park ($). There's a ramp and kayak/canoe rentals, and you can also launch along the road. The typical freshwater fish species reside in the lake. Word is that the fishing there is not that good.

That dedicated, motivated paddler can start a downriver trip from here. It's about a three-day paddle to El Jobean, but fishing will slow you down significantly. You won't find many camping places. Below State Road 72, the park requires you to have a wilderness permit. The alligators in the lower lake are large and plentiful. Large alligators occur all along this stream. We still have all the freshwater fish species here.

You'll find a kayak launch at Venice Myakka River Park in Nokomis. The river now has less wilderness character, but you start encountering saltwater

Upper Deer Prairie Creek, a lovely place.

fish, especially snook. If you don't mind the road and construction noise, take a full day to float the ten miles from here to Snook Haven; this old Florida-style fish camp on the Myakka River is definitely worth a visit. Snook Haven has rental canoes but does not offer a shuttle. You can get a shrimp po'boy or some barbecue—a perfect cap to a long day's paddle. You can even buy cold, carbonated malt beverages!

Another, shorter float trip runs from Snook Haven to Deer Prairie Creek Preserve, about seven miles. You'll still catch both fresh- and saltwater fish through here. You'll be paddling through people's backyards, will have lots of road noise and motor vessels and may have to paddle upstream about a mile against a slow current once you find the mouth of Deer Prairie Creek on the left side of the river just before reaching the Highway 41 bridge. The kayak launch will be on the right after about a mile. If you get to the dam, you missed it!

You could also paddle the remaining nine miles from Highway 41 to El Jobean. Fish species will all be saltwater types here. The river is fairly wide, and the motorboat traffic is unacceptably heavy, so paddling this stretch is not recommended.

DEER PRAIRIE CREEK

A small dam on Deer Prairie Creek backs up a small pond. Fish above the dam if you want freshwater fish. A small creek feeds the pond and can be paddled.

This creek is intimate and gorgeous. As the morning passes, the songs of birds gave way to the humming of cicadas. If you listen, you can hear traffic sounds, but they are distant and easy to ignore. Hawks scream. Owls hoot.

If you go too far, turning around will be difficult. The stream is too overgrown in most places to fly fish. An ultralight spin outfit will work if you're careful.

On my last visit, a four-weight fly rod did come in handy, though. On the return trip, once the creek started opening up, the fish started hitting my popper. The first was a feisty little bass, quickly followed by a garfish. After five bass and a stumpknocker, the sun got too high and hot for me to continue. Glad I had found this tiny gem, I loaded the boat back onto my car.

Below the dam, the creek is still attractive, and you can find snook and sometimes some other saltwater fish. The flow reverses in this creek, so pay attention to the El Jobean tides. Do the trip as an out-and-back. It's a touch more than a mile to the main channel of the Myakka River, which also bears some investigation. Expect to see traffic from motor vessels.

OTHER CHARLOTTE COUNTY PADDLE FISHING

A creative paddler could fish Charlotte County for a month and never hit the same water twice. There are natural ponds, retention ponds, creeks, residential canals—all kinds of water. Fish range from native species like bass, snook and tarpon to exotic species like cichlids, peacock bass and snakeheads.

One morning, I met kayak-fishing guide Logan Totten (www.flyakker.com) in Englewood. In the dark, he drove me somewhere into the Charlotte Harbor Preserve State Park. The sun was not yet up when we launched the kayaks into a small pond.

After crossing the pond, we dragged the boats into a second pond and began fishing. Baby tarpon rolled around us. In spite of several fly changes, baby tarpon did not bite. A few Mayan cichlids did, so no skunking for us!

Kayak guide Logan Totten drags his boat from one pond to another.

Logan led me through a mangrove tunnel (I love mangrove tunnels!) into another pond. Again, rolling tarpon, no biters.

We eventually went back through the tunnel into the second pond. We got more Mayan cichlids and a few small snook. The tarpon still refused to cooperate.

We dragged the boats back to the first pond, paddled back to his truck, loaded up and went to another pond. Tarpon rolled all over it. We took two drifts across without a bite and called it. Tarpon of any size can be so ornery…

Totten works out of the West Wall Fly Shop in Port Charlotte, where they sell kayaks and accessories in addition to their lines of fly tackle. It's a great resource if you're in this part of Florida.

Logan Totten casts to rolling baby tarpon in a pond.

Nuts and Bolts

Launch or recover your vessel at:

- Myakka River State Park ($), rentals and camping available
- Venice Myakka River Park in Nokomis
- Snook Haven, rentals, food, and drink available
- Deer Prairie Creek Preserve; no shuttles are available on the Myakka or its tributaries—you are on your own

Josh and Logan at West Wall Fly Shop (941.875.9630) offer fly tackle, kayaks and accessories, guided trips and lots of information about paddle fishing this area.

The Charlotte County Visitor's Bureau is at www.charlotteharbortravel.com.

THE MYAKKA RIVER AND THE FOUNTAIN OF YOUTH

According to historical documentation by Antonio De Herrera, Juan Ponce de León may have been the first European to explore the Charlotte Harbor area. Herrera was the official historiographer of the Indies, appointed in 1592 by King Phillip II of Spain. It is believed that Herrera possessed Ponce de León's logbook.

In 1511, Ponce de León obtained the permission of King Phillip II to explore the mysterious lands north of the Indies. Herrera states that the explorer reached the North American continent in April 1513 in the area of the mouth of the St. Johns River on the east coast. Herrera records the origin of the name "Florida" using words that may be those of Ponce de León himself. A translation of Herrera's writing: "They named it La Florida because it had a 'very beautiful view of many and cool woodlands and it was level and uniform,' and because moreover, they discovered it in the time of the Feast of Flowers (Pascua Florida)."

Ponce de León proceeded south along the east coast of Florida. He searched for an area on the west coast where the Indians reportedly had gold. By May 23, Ponce de León's expedition was in the yet unnamed Charlotte Harbor area. Here, according to Herrera, "he found the passage for vessels next to the coast into Carlos Sound and anchored off Pine Island…in this haven he careened the San Christoval, and later sent a boat to examine and sound a harbor nearby." Herrera was certain that Ponce de León "examined the region with a thought of future settlement." With this in mind, it can be assumed that one of the exploratory scouting boats may have encountered the two major rivers flowing into the harbor. These rivers that flow into Charlotte Harbor are the Peace and the Myakka.

Herrera records that Ponce de León was not only in search of gold but also "went seeking that sacred fountain, so renowned among the Indians, and the river, whose waters rejuvenated the aged." This is among the earliest records mentioning the "Fountain of Youth" in Florida. Interestingly, Herrera also mentions the river that shared the fountain's rejuvenating effect. If we assume that Warm Mineral Spring is the ancient Fountain of Youth, then it can also be assumed that the Myakka River, ultimately accepting the water flowing from the spring, is the legendary river.

You can still visit the Warm Mineral Spring (www.cityofnorthport.com/visitors/visit-north-port/warm-mineral-springs-park)—if you're willing to part with twenty dollars. The park is in North Port. You can get a massage and a facial there, too—something Ponce de León likely would not have received.

Ponce de León remained in the Charlotte Harbor area until June 15, 1513. He then returned to the West Indies and prepared for his second journey to "La Florida." Since Ponce de León had considered the Charlotte Harbor area suitable for a settlement, the region was likely the destination of his second voyage.

Herrera recorded that second voyage. The attempt at settlement was foiled by the native Indians, who "sallied out to oppose" Ponce de León. A battle ensued, and many members of Ponce de León's crew were killed, and many others were wounded. Ponce de León, himself wounded, returned to Cuba, where he died as a result of the battle.

PEACE RIVER

Similar to the nearby Myakka, the Peace River originates from two tiny creeks that join north of Bartow—Peace Creek and Saddle Creek, although many people consider Hancock Lake to be the start. Either way, the river runs south for over one hundred miles before emptying its waters into the top of Charlotte Harbor.

Above Fort Meade, the river runs shallow and often has little or no water during the dry season. For paddling anglers, the best stretch runs from Fort Meade to Arcadia, which is about fifty miles if you paddle the length. You'll need to camp! Overnight trips here are delightful, but you can also take day trips. Most require a shuttle unless you want to do an out-and-back against a sometimes considerable current.

Fish species include bass, sunfish, catfish, snook and, rarely, a redfish. Because of those snook, you may want to use beefier tackle and heavier leaders than you otherwise might.

Downstream from Arcadia, the river is navigable and used fairly heavily by power vessels of all sizes. If this doesn't bother you, by all means, try it—there are plenty of saltwater species there. This stretch is tidally influenced, and you'll have to pay attention to the tide chart.

The river in the recommended stretch has lots of shallow sandbars and several limestone shoals. Available campsites (not many!) are all on the west bank; the east bank is private property. Another interesting note about the river—the bottom is loaded with fossils. If you spend some time sifting through some mud, the chances of finding fossilized shark's teeth (among other things) are quite high.

On our recent trip there, Mike Conneen and I launched our boats at the Gardner boat ramp, intending to take-out the next day in Arcadia. A friend helped us with the shuttle. Then, we were off, pushed along by a lively current.

Destinations

The cypress roots along the Peace River are fantastic!

Mike used a spinnerbait to pull this Peace River snook out of a logjam.

We found fishing slow, but our trip coincided with the coldest weather of the season. We did catch fish—a few sunfish, a few small bass, and Mike got a snook out of a logjam when the fish struck his spinnerbait. My guess is that fishing improves when the temperatures are warmer.

Nuts and Bolts

Rentals are available at Peace River Canoe Outpost (800.268.0083, www.canoeoutpost.com/peace).

These folks have somewhat of a monopoly on the Peace River, providing boats, shuttles, information and campsites. It's a one-stop shop!

History

Today, the Peace River is an economically unimportant, small and winding river in central Florida; the upper reaches of the river shrink in the dry winter months to a mere trickle of water. Looking at the river now, it is hard to believe that the Peace was once a frequently traveled route that connected central Florida with the Gulf of Mexico. Before the advent of modern transportation networks, the Peace River functioned as a highway of sorts.

Early natives, especially the Calusa, used the Peace River system up into the Green Swamp, north of present-day Lakeland, as a passageway for their dugout canoes. The Seminoles also used the river for transportation. Because of the natives' reliance on the river as a means of travel, it follows that when the soldiers of the Third Seminole War moved into the area, they, too, used the river to find the elusive Seminoles.

Later, as the threat of Indian attacks subsided, settlers began moving into the area, which, ironically, led to the disuse of the Peace River. As the population grew, the building of roads and railroads brought more reliable and direct means of traveling, making river transportation less important. Like most other Florida rivers, the Peace River was eventually forgotten as a transportation route. Today, only those who enjoy fishing and canoeing have seen this beautiful stream—except when crossing its bridges in their cars.

As early as 1000 BC, natives used the Peace River to travel between the Green Swamp and the Gulf of Mexico. Their shallow draught dugouts would have had no difficulty traversing the river even during the lowest stages of water.

Destinations

The late Park DeVane, a longtime student of the Seminoles, has an interesting theory concerning the Seminoles' use of Florida rivers. According to DeVane, it was once possible—by using the Green Swamp in the fashion of a railroad roundhouse—to travel from one part of Florida to another via the Peace, Kissimmee, Alafia, Hillsborough, Withlacoochee, Oklawaha, St. Johns and Econlockhatchee Rivers. All of these rivers had their headwaters in the Green Swamp or connected to another river that did, so it was possible to travel through the swamp to get from one river to the next. Using this, the Indians could have traveled continuously by water from the Gulf of Mexico to the Atlantic.

DeVane based this theory on the scores of years he spent interviewing those Seminoles still living. He learned that the Indians once traveled practically anywhere in Florida in their canoes. Using linking lakes, creeks, rivers and inundated swamps, it was possible to travel long distances. DeVane published an account by Billy Bowlegs III of a canoe trip from Fort Basinger, on the Kissimmee River; through Lake Okeechobee; up Fisheating Creek to Rainey Slough and Gannett Slough; then into Myrtle Creek, Shell Creek and up the Peace River to Fort Ogden, where he traded and then made the return trip. The 260-mile round trip does not seem possible now that the smaller bodies of water are dried up, but this was once common practice.

In 1880, Congress passed an ambitious act calling for a cross-Florida steamboat canal. The waterway was to connect the St. Johns River to the Peace River by way of the Topokalija Lake without the use of locks. The river was surveyed from Fort Meade to its mouth as a part of the Army Corps of Engineers examination of the canal route. As a result of the surveys, the Corps of Engineers concluded that a steamboat canal along this route could not be built without the use of locks, the cost of which far outweighed the benefits to a sparse Florida population. Fortunately, the canal idea died.

Unfortunately, the survey reported the presence of phosphate in the river, and businessmen soon organized phosphate hunts. Working in secret so as not to drive up land prices, these men and their scientists floated down the river taking random samples of the riverbed. When they were satisfied that huge amounts of phosphate could be recovered from the river, they quickly set to work.

By 1891, the fledgling phosphate industry had bought up acreage on both sides of the river and begun mining the river bottom. Dredge barges, which resembled large rafts with room aboard for a steam engine and a centrifugal pump, littered the river. Accompanying the large barges were smaller "lighters" that conveyed ore to drying houses on the riverbank. Once the phosphate was dried, the lighters hauled it downriver to Punta Gorda, where oceangoing steamships carried it to domestic and foreign ports.

Within a short time, mining was so intensive that the phosphate industry had practically taken over the river.

The miners eventually became exasperated with using the river for transporting their vessels. During the dry months, low stages of water were often too low for floating their laden boats, and high water in the rainy season sometimes flooded their works.

The miners welcomed the coming of the railroad, which promised a more reliable means of getting the phosphate to Punta Gorda. As the companies began abandoning river mining altogether for more profitable land mining, there was even more reason to rely on the railroad for the transportation of phosphate to Punta Gorda. The Peace River was soon abandoned in terrible condition. The river was now cluttered with sunken barges, fallen trees and new shoals created by the dredging operations.

Today, following the river by canoe from Bartow, one encounters few other boaters but many reminders of past activity. Just north of Fort Meade, there is a broad place in the river where the Seminoles dammed the water to make a fish weir. One can still see the evidence of a long-gone cypress-logging industry in the upper river. Near Bartow is a sunken phosphate dredging barge, the outline of which can be discerned at low water. The remnants of wharves built by the earliest phosphate companies still stand near Arcadia. These physical remainders remind one that this river was not always as quiet as it is today. The Peace River, once a frequently traveled route, is now a forgotten highway.

PEACOCK BASS IN SOUTH FLORIDA

Florida's Fish and Wildlife Conservation Commission (FWC) introduced peacock bass to southeast Florida's canal system in 1984. Since this is all fishing through backyards and such, I wouldn't suggest a special trip to fish for them. But, if you find yourself in Broward or Dade County, it might be fun to try.

Peacocks prefer live fish and the fish-imitating baits often used by largemouth bass anglers, but they rarely hit plastic worms commonly used to catch largemouth bass.

A peacock bass caught on a fly in a Broward County canal.

The average south Florida peacock runs between one and three pounds. You can catch them throughout the year, with most of the larger specimens, up to ten pounds, caught between February and May. They can be sight fished then—of particular interest to fly-fishers.

Shaded areas provided by bridges, culverts, docks and other structures generally are productive fishing spots, along with fallen trees, canal ends, bends and intersections. Most peacocks are caught during daylight hours.

Topwater lures such as Heddon's Baby Torpedo and Storm's Chug Bug work well. Work these baits fast. Be aggressive! A lot of bass anglers work the baits too slowly for peacocks.

Minnow-imitating crankbaits, like the three-inch Rapala, also work well. Yellow, gold and orange baits show up well in the often discolored water of the canal systems.

Jigs fished on casting or spinning tackle are good choices for artificial baits. Small tube lures and jigs frequently are used to sight fish peacocks, especially when they are aggressively guarding spawning beds near the shoreline.

Although bigger baits may entice more trophy-sized fish, baits less than three inches in length will produce more consistently than larger ones. However, even big peacocks will take baits smaller than largemouth bass anglers typically use.

Fly casters can use flies such as divers, deceivers, EP-style streamers, Clouser Minnows and poppers with success. Try chartreuse or yellow flies with flashy strips of Mylar-type materials. Flies should be stripped rapidly. If you aren't working hard, don't expect a lot of bites.

Cast flies close to the shoreline. Pay attention to fallen trees, docks or any other structures. Take three or four fast strips, then make another cast. Sinking lines are preferred for this work.

Spin fishermen should use light tackle with six- to eight-pound test line. Light lines and tippets generate more strikes than heavier ones, and heavier lines aren't necessary, because canal-caught butterfly peacock tend to be open-water fighters.

A peacock bass can be handled by its lower jaw using the same thumb-and-finger grip one would utilize for largemouth bass, although this will not immobilize them. By the end of the day, successful anglers using this grip will have many minor thumb scrapes caused by the peacock's sandpaper-like teeth.

The FWC encourages anglers to practice catch-and-release when fishing for butterfly peacock. Overall, this species is a hearty fish, and nearly 100 percent will survive being caught and released when properly handled. However, butterfly peacock do not survive as well in live wells or as long out of water as largemouth bass. It is important that they be quickly released to maximize their chances for survival.

There are 1,100 miles of canals in peacock country, 400 or so of which support peacocks. The uninitiated will need some help zeroing in on the more productive areas.

The best and most up-to-date fishing reports for peacock bass are available from local bait-and-tackle shops. For first-time, nonguided butterfly peacock anglers, it is strongly recommended to check with local freshwater tackle shops for the best locations and baits to use. However, quarterly fishing forecasts are also available on the FWC website at myfwc.com/fishing/freshwater/sites-forecasts/s/metropolitan-southeast-florida-canals.

The following Dade and Broward County canals get high marks for peacock bass: Tamiami Canal (C-4), Biscayne Canal (C-8), Cypress Creek Canal (C-14), Cutler Drain Canal (C-100), Snake Creek Canal (C-9) and Snapper Creek Canal (C-2). Those looking for quality largemouth bass *and* butterfly peacocks might try the Cypress Creek Canal (C-14), Cutler Drain Canal (C-100), Airport Lakes Canal (C-4) and Biscayne Canal (C-8).

Typically, these canals are kept weed-free, giving light-tackle anglers a reduced risk of hang-ups and a greater opportunity to land a wall-hanger. Phenomenal peacock and largemouth fishing activity now awaits anybody with a hankering to do battle with world-class finny adversaries.

The Everglades Wildlife Management Area (myfwc.com/recreation/lead/everglades), west of Fort Lauderdale, has both largemouth and peacock bass available in its canals.

For more information, contact the FWC Everglades Regional office at 561.625.5122 or visit floridaconservation.org/fishing/offices/westpalm.html.

PART 4
PADDLE FISHING THE SEA

The Atlantic washes the east side of Florida, the Gulf of Mexico caresses the west. You can find good fishing off public access beaches on both coasts if you're careful and plan your trip with care. Here are a few suggestions.

1
PADDLE FISHING FROM THE BEACH

It's a manly thing to do, launching your paddle vessel in the surf and paddling out onto the Atlantic or Gulf to search for the bounty of the sea. I don't do this often, but I have done it enough to give the potential ocean paddler a few tips.

Anyone who wants to try this should first make a dry run without tackle. Take your vessel—be it canoe, kayak or paddleboard—to the beach, then launch it and beach it again several times without anything of value in the boat. Wear your life jacket when you do this. There's a learning curve here, and if you dump while figuring things out, you don't want to lose your tackle. Actually, you should wear your life jacket every time you launch from the beach. It's not like flats fishing, where you're in inches of water—this is the ocean. It can be unforgiving of errors.

You will want to try these launches and beachings several times. You also may want to intentionally dump so you can learn self-rescue. When you feel like you've got the knack of making it past the breakers with élan and panache, when you feel confident you can handle yourself if you have a problem, then it's time to get the tackle and try it for real.

You'll find when you try this that fishing canoes and kayaks don't surf well. The proper technique for dealing with breakers when you're coming back in is to backpaddle through them rather than trying to ride them. In my experience, trying to ride any waves more than six inches high in these boats results in the boat quickly turning sideways and rolling over. Leashing valuable rods and reels, as well as your paddle, is not a bad idea at all. Taking your less-valuable equipment through the surf is not a bad idea, either.

If you intend to paddle fish in the ocean, it makes sense to avoid the high-traffic motorboat areas. Stay away from places where there are a lot of motorboats! While a paddler has the right of way over a motor vessel, it's safer to not trust motor-vessel operators to be familiar with the rules.

Before you go, you'll have to decide what your target fish will be. You'll need different tackle for Spanish mackerel, whiting, pompano or bluefish than for tarpon or big sharks. You don't have space for a wide variety of stuff, and there's always the risk of losing it in the surf, so put some thought into it. My suggestion would be to start with the smaller fish species, and once you're familiar with the entire drill, work your way up to the big boys.

Before you launch, scan the water for signs of activity. Bait pods are always good to see. If fish are busting them, get out there immediately! Rolling tarpon and diving pelicans are other wonderful things to see. If you don't see anything, consider looking somewhere else. You can't catch fish where there are none.

I haven't done this, but you can catch tarpon off the beach from paddle vessels. Some folks toss or troll lures. Others use bait. If you opt for live bait, you're going to have to keep it alive. A half a ladyfish or mullet makes more sense than worrying about a live well system, although such systems are available, if that's the way you must go. Swimbaits or the DOA BFL are great beach lures for tarpon. Sharks will just as readily eat a half a dead ladyfish as any live bait.

Birds work over breaking fish off a Florida beach—a wonderful thing to find!

Any smaller fish you hook can just be reeled in. Big fish require technique. If possible, turn the boat at right angles to the fish to increase the resistance it's working against. A big tarpon can easily pull around any small vessel for a long time. You want to shorten the fight as much as you can.

While I enjoy fishing alone, this is not the place to do that. Fish with a buddy, and make sure someone on shore knows the area(s) where you intend to fish and when you're expected to return. Call that person just before you launch so there's no ambiguity.

Launching off the beach into the ocean from a paddle vessel is always a grand adventure. I wish I lived closer to the beach so I could do it more often!

2
KAYAK FISHING OFFSHORE

BY MIKE CONNEEN

With Florida's 1,350 miles of coastline, there's no shortage of places to launch a paddle or pedal craft into either the Atlantic Ocean or Gulf of Mexico. In these ocean waters, you will find fish that are going to fight harder than any species of inshore or freshwater fish.

It's likely that the first time a dugout canoe was crafted, the natives became curious about deeper waters after they finished exploring local waters. Create some outriggers to enhance the stability of the vessel, and you have a big-water fishing machine.

Centuries later, with a splash of engineering and a dose of polyethylene, you now have what are infinitely adaptable plastic kayaks. Some of today's kayaks are designed for big-water fishing. The stability of these small boats is similar to standing on a floating dock. They can penetrate large breaking waves, popping out of the backside like an arrow.

You must develop trust in your vessel before going out to sea, because your life depends on that boat. Going into the ocean with an untrustworthy kayak can cost you your life. A comfortable and reliable PFD (personal floatation device) should be worn at all times, regardless of swimming skills.

In addition to a PFD, sporting a brightly colored, easily visible flag that stands tall and carrying a waterproof, handheld VHF radio is recommended. Brightly colored kayaks might not be the hippest thing to paddle, but they can save your life and are likewise recommended. If you are in a naturally colored kayak and wearing light colors, it will be very hard for large yachts and sportfishing boats to see you.

Large swells sometimes occur. They will drastically raise and lower your craft. Each time you are lowered into the trough of a swell, you become invisible to other vessels. Swells can also crest, and you should always be aware of your surroundings, because a wave caught when you are off-guard will flip you over before you know you're wet. Instructional videos on kayak reentry in deep water can be found online and should be studied prior to going out for your first time.

The ocean is unpredictable and shows no mercy, so always double-check that all of your gear is in good condition before every trip. The time you spend doing this is an investment in additional security. Safety is always the number-one priority, because if it's any lower on the prep list, you may not return.

Finally, this is not the place for rugged individualism. Going out to sea in a tiny boat entails risk. Having a friend or two along mitigates some of that risk. Deepwater kayak fishing should always be done with one or more competent friends.

"Beyond the breakers" is the term used for fishing outside of the breaking waves. Launching into the surf takes practice and skill. You need to have plenty of both.

Any gear you have that is exposed should be secured to the kayak. Rod and paddle leashes are a wise investment. I have had a bad experience with the surf—it ruined my day.

Pick a day with small waves to practice paddling out and in. Going out will always be easier. You simply aim your kayak directly into the waves and

A big king mackerel caught well off the beach. Note that the fisher is wearing a personal flotation device (PFD), and the rod is leashed to the boat. *Courtesy of Brian Nelli.*

paddle. This is best done with a sit-on-top kayak with open scupper holes, so the water can quickly drain from the boat before the next wave comes.

Coming back in is the hard part. Waves can be ridden in on the back side, with the kayak pointed toward the beach. This takes good timing, but waves typically come in sets. Wait for the waves to build up, and when they start to dwindle, ride the back of the small waves in before they break.

If you get caught on the face of the wave, aim diagonally across the wave and dig your paddle deep to surf the wave. The most common place for people to lose gear and flip kayaks is in knee-deep water while coming in.

Geographically, most of the Florida coast has a shallow shelf next to shore, with the exception of the southeast region. This coast, from central to south Florida, is where you will find the most difficult waves to traverse. However, the southeast coast is by far my favorite place to kayak fish.

If you want a large diversity of large fish, find your way here. This area stretches from Pompano Beach south to Miami. The diversity of fish—tarpon, sailfish, king mackerel, giant jacks, mahi, sharks, various grouper and snapper species—is what draws me to these waters. The Gulf Stream flows through here and is within paddling distance from shore. You can fish in a couple hundred feet of water and still be reasonably close to land.

There's a distinct color change in the Gulf Stream that you won't find anywhere else around the state. Hovering in a small plastic boat on top of beautiful sparkling aqua-blue abyss, miles from shore, you will feel very small. Sea life will visit you and help you gain a little more appreciation for this great body of water. You can troll live or artificial baits on top or lower them with a downrigger.

When using live bait offshore, it's best to use a circular bait well with an aerator so your bait can swim in a continuous circle. This will reduce the stress on the bait and prolong its life.

The biggest difficulty in this area can be getting your gear across the sand. A kayak cart is a must, especially if you have a large kayak. You can launch from any public beach access. Most will have a parking fee.

In this area, the Gulf Stream hugs the coast and gives kayak anglers an easy opportunity to paddle into four hundred feet of water with little effort. However, the Gulf Stream can move very quickly at times, and it's always important to keep landmarks in sight. Pedal-propelled kayaks dominate this type of fishing due to the fact that the current is strong. Having a pedal craft is a bonus when fishing structures such as wrecks, because you can fish a small window over the structure without losing position. Common species caught in this manner can include tuna, sailfish, amberjack and wahoo, but the list is much longer due to the habitat that these fish enjoy.

On the other side of the state, the Gulf of Mexico hosts thousands of operational and abandoned oil platforms. These giant structures in the water create habitats for some great fish. Unfortunately, they are not easily accessible by kayak. The best way to reach these platforms with a kayak is to hire a "mothership" (find these motherships via internet search, for example, "kayak mothership Pensacola") because of how far they are located from shore. The mothership can transport you and your kayak to these spectacular fishing grounds.

When you hook into a significant fish in open water, you will have a fight of a lifetime. You cannot appreciate the true strength in some of our ocean's fish from a large boat. When you're in a large motor vessel, the resistance that fish must fight is thousands of pounds heavier than that of an angler in a two-hundred-pound boat. The kayak gives the fish a much better chance of getting away, because it's able to pull you and your rig out into the middle of the ocean.

When fishing offshore, there are two types of fights you will experience. A large tarpon, kingfish or sailfish is going to take line from you in a horizontal direction. This allows you to control the fish more due to the lack of objects high in the water column. You're less likely to lose that fish due to the line breaking. Don't be surprised when they pull you for miles.

A structure-oriented fish, such as a grouper or snapper, will give you a vertical fight. Structure fish are difficult to boat if you do not get the fish away from the structure on the seafloor. These structures can be made of coral, rock ledges, pieces of concrete, sunken ships and other objects—the coordinates of many artificial reefs, including sunken boats, can be found on the internet. This is where structure fish live, and they utilize their environment for protection.

Once the fish is hooked, it will try to go back to the structure and break you off, leaving your tackle at the bottom of the ocean. The initial fight requires you to lift the fish away from the structure and keep that fish from going back down. You must get the fish away from the bottom before you focus on landing it. Once the fish is landed, you will typically be in the vicinity of where you hooked it.

Bottom fishing is most commonly done with live bait or by dropping vertical jigs down. Vertical jigging is my favorite type of fishing, because it's a workout full of adrenaline. I find jigging much easier from a kayak than from a full-sized boat. While jigging, the kayak looks and sounds more like floating debris than a motorboat, so you can expect curious fish to pay a visit. Carry a lighter rod for casting at fish, like cobia or mahi-mahi, that come to investigate.

When bottom fishing for sport, be sure to properly ventilate the fish before release if it shows any signs of barotrauma. Ventilation reduces post-release mortality. Ventilating needles can be found at local tackle stores and online. Instructions for their use can be found online. YouTube has some good videos on this subject.

Whether you're fishing high in the water column or on the bottom, fighting large fish in the ocean from a kayak is extremely exhilarating. Keep in mind that safety must be heavily considered while on open water. Handling big fish in big waters can be dangerous, and as mentioned earlier, the buddy system is highly recommended.

A big fish can severely injure you. They could weigh well over one hundred pounds, and many have teeth and spines. If the fish is not exhausted when you boat it, it could get wild on you—quickly. Using a dehooking device while the fish is in the water is a very safe way to unhook fish you intend to release.

For fish coming aboard for consumption, subdue the fish with a club or by bleeding it prior to pulling it onto your lap. A live fish on ice will eventually die—but not without stress and possibly ruining the meat. Euthanizing these fish will not only protect the meat, but it shows respect for the great resource we have.

Be sure your bail is always open when handling fish until the hook is removed. Combine proper dehooking and a rod leash, and the chances of losing your rod to the sea significantly decrease.

Kayak fish-cooler bags can be found online and work well for carrying any fish you intend to take home and eat. I personally carry one, made of a reflective material, on the bow of my kayak. In addition to carrying my fish, it creates an easily visible point that helps other boaters to be able to see me.

Our oceans are a beautiful place. Unfortunately, they have become a litter bin for our planet. Two things we need to survive in this world are clean air and clean water. Every time I fish in the ocean, I bring something to collect the plastics and other floating trash I find. In this small way, I thank the ocean for giving me the opportunity to enjoy its endless wonders. Hopefully, you will do the same, leaving the place cleaner than how you found it.

Be safe, go fishing, take photographs and make memories.

APPENDIX

FLORIDA FISHING LICENSES

From the FWC website: "Florida residents and visitors need a Florida freshwater or Florida saltwater fishing license unless they are a member of one of the exempted groups of people listed below. Your license is required to be with you when you are engaged in the licensed activity."

The rules are too complicated and change too often to get into here. Visit myfwc.com/license/recreational for up-to-date information.

The easiest way to get a license is to get it online at gooutdoorsflorida.com or get it over the telephone by dialing 888.347.4356.

FOR MORE INFORMATION

To find more information about areas you want to explore, I recommend the following resources:

- The Fish Finder (an excellent resource for Citrus County from Bayport to Yankeetown): fishingkayaks.us/launch-sites
- Florida's Designated Paddling Trails (all listed in one place): floridadep.gov/parks/ogt/content/floridas-designated-paddling-trails

Appendix

- Google Maps (I almost can't remember how we lived without this): maps.google.com
- Nature Coast Kayak Launches: fishingkayaks.us/launch-sites
- Paddle Florida Trip Descriptions (includes one hundred or so paddling trips): www.paddleflorida.net/index.htm
- Southeast Florida Canal System (fish for bass, peacock bass and a variety of other exotic species): myfwc.com/fishing/freshwater/sites-forecasts/s/metropolitan-southeast-florida-canals
- Timicuan Trail (Jacksonville) Paddle Guide: timucuantrailwaterwayguide.org

GEAR LISTS

You may not need all this stuff on your trip, but it's nice to have access to a comprehensive list. Modify these lists to suit your needs!

Boat

Canoe/kayak/paddleboard
Stake-out pole/anchor
Small tool/repair box
Paddles—one per person, plus a spare
Personal flotation device (PFD)—one per person
Whistle
Bow/stern lines

Fishing

Fly and/or spin rods
Small conventional tackle box
Fly bag
Leader materials

Appendix

Housing

Tent
Ground cloth
Sleeping mat
Sleeping bag
Pillow (small)
Tarp
Pee bottle

Clothes

Sneakers
Sandals
Baggy light pants (at least three pairs)
Shorts (at least one pair)
Jerseys (at least four)
Underwear (as needed)
Belt
Socks (wool is best, as needed)
Warm long-sleeved shirt
Hoodie
Rain gear
Buff
Hat
Sunglasses
Flats booties or waders and wading shoes

Personal

Towel(s)
Toilet tissue
Pocketknife
Dental care items
Small flashlight
Eye/sunglasses
Camera

Appendix

Notebook and pencil
Soap/shampoo
Repair kit (needle and thread, duct tape)
Phone and solar charger
Prescription medicines

Kitchen

Stove and fuel
Matches/lighter
Can opener
Pots (small, medium, large, with handle)
Towel/potholder
Colander
Fry pan
Salt, pepper, spices
Eating utensils
Dish soap, sponge, towel(s)
Food containers
Cooler

Miscellaneous

Waterproof pack
First-aid kit in waterproof bag
Maps and compass
Nylon cord
Ziplock bags
Small trash bags
Votive candles
Matches
Sunscreen
Bug spray

BIBLIOGRAPHY

Apte, Stu. *Stu Apte's Fishing in the Florida Keys and Flamingo*. Miami: Windward Publishing Company, 1976.
Belleville, Bill. *River of Lakes: A Journey on Florida's St. Johns River*. Athens: University of Georgia Press, 2001.
Derr, Mark. *Some Kind of Paradise*. Gainesville: University Press of Florida, 1998.
Dimock, Anthony Weston. *The Book of the Tarpon*. Outing Publishing Company, 1911.
Dunaway, Vic. *Sport Fish of Florida*. Stuart: Florida Sportsman, 1998.
Florida Department of Environmental Protection, Surface Water Monitoring Section. *Current Biological Health and Water Quality of the Econlockhatchee River and Selected Tributaries, April 2000*. Orlando: Florida Department of Environmental Protection, 2000.
Francis, Phil. *Florida Fish and Fishing*. New York: Macmillan, 1955.
Grunwald, Michael. *The Swamp: The Everglades, Florida, and the Politics of Paradise*. New York: Simon & Schuster, 2006.
Huff, Sandy. *Paddler's Guide to the Sunshine State*. Gainesville: University Press of Florida, 2001.
Kreh, Lefty. *Fly Fishing in Salt Water* (revised edition). Guilford, CT: Lyons Press, 1986.
Kumiski, John. *Fishing Florida's Space Coast*. Chuluota, FL: Argonaut Publishing, 2003.
———. *Fishing the Everglades*. Chuluota, FL: Argonaut Publishing, 1993.

Bibliography

———. *Flyrodding Florida Salt*. Chuluota, FL: Argonaut Publishing, 1995.
Matthiessen, Peter. *Killing Mister Watson*. New York: Vintage, 1991.
———. *Lost Man's River*. New York: Vintage, 1998.
———. *Shadow Country*. New York: Modern Library, 2008.
McGowan, Bob, and Richard Farren. *Fishing the Big Bend*. Tallahassee, FL: Woodland Productions, 1993.
Milanich, Jerald T. *Florida's Indians and the Invasion from Europe*. Gainesville: University Press of Florida, 1995.
———. *Florida's Indians from Ancient Times to the Present*. Gainesville: University Press of Florida, 1998.
Sarasota County Historical Archives Staff. *A History of the Myakka River, Sarasota County, Florida*. Self-published, 1983.
Sosin, Mark, and Lefty Kreh. *Fishing the Flats*. Guilford, CT: Lyons Press, 1983.
St. Johns River Water Management District. *Recreation Guide to District Lands, Fifth Edition*. Self-published, 2006.
Tebeau, Charlton W. *Man in the Everglades*. Miami: University of Miami Press, 1968.
Thompson, Tommy L. *The Saltwater Angler's Guide to Florida's Big Bend and Emerald Coast*. Gainesville: University Press of Florida, 2009.
Urban, John T. *A Whitewater Handbook for Canoe and Kayak*. Boston: Appalachian Mountain Club, 1976.
Ware, Lynn W. *The Peace River: A Forgotten Highway*. Tampa: University of South Florida, 1984.
Wenner, Dr. Charles. *Red Drum Natural History and Fishing Techniques in South Carolina*, Educational Report No. 17. Columbia: South Carolina Department of Natural Resources, 2005.

ABOUT THE AUTHOR

Courtesy of Mike Conneen

Fishing is in John Kumiski's blood. Blame it on his father, who got John started fishing while he still wore diapers. He's been tying flies since he was a child and has caught seventy species of fish in sixteen states and ten countries. His first paddling experience involved a wood-and-canvas Old Town canoe on Little Sebago Lake in Maine—he was eleven years old.

John has served in the U.S. Army and holds a bachelor's degree from the University of Massachusetts. Long ago and far away, he labored as a public school teacher. For years, with the help of the Backcounty Flyfishing Association and the Orlando Kiwanis Club, he organized an annual Fishing Day for Kids.

John specializes in sight fishing with fly and light tackle for any fish that will take a fly or lure. John has taught fishing classes at Brevard Community College and was the instructor at the Andy Thornall Fly Fishing for Redfish School. The FFF (Federation of Fly Fishers; now called Fly Fishers International) has certified John as a casting instructor. He frequently speaks to fishing clubs, gives seminars and performs similar activities.

John is a member of the Southeastern Outdoor Press Association. He has been a three-term president of the Indian River Guides Association, a two-term president of the Backcountry Flyfishing Association and president of

About the Author

the Florida Outdoor Writers Association. He was selected as a "Top Rated Guide" in *Top Rated Saltwater Fishing: Bays, Estuaries, Flats & Offshore in North America* by Maurizio Valerio (Lanham, MD: Derrydale Press, 2000).

In addition to guiding and writing books, John writes freelance magazine articles and indulges in photography. His most recent books include two about his beloved Space Coast area; they are titled, appropriately enough, *The Indian River Lagoon Chronicles* and *Fishing Florida's Space Coast*. He also authored *Redfish on the Fly* and *Flyrodding Florida Salt*.

Eight boats of various types call John's yard home. John is always ready to talk fishing and can be reached via his websites: www.spottedtail.com and www.johnkumiski.com.

www.ingramcontent.com/pod-product-compliance
Lightning Source LLC
Chambersburg PA
CBHW040303170426
43194CB00021B/2873